CATASTROPHES AND CONFLICTS: SCIENTIFIC APPROACHES TO THEIR CONTROL

To my dear wife Karin

Catastrophes and Conflicts: Scientific Approaches to Their Control

KLAUS GOTTSTEIN
Max-Planck-Institut für Physik
Munich, Germany

Ashgate

Aldershot • Brookfield USA • Singapore • Sydney

Published by
Ashgate Publishing Limited
Gower House
Croft Road
Aldershot
Hampshire GU11 3HR
England

Ashgate Publishing Company
Old Post Road
Brookfield
Vermont 05036
USA

Ashgate website: http://www.ashgate.com

British Library Cataloguing in Publication Data
Gottstein, Klaus
 Catastrophes and conflicts: scientific approaches to their
 control
 1.Industrial accidents 2.Industrial safety 3.International
 relations
 I.Title
 363.1'1

Library of Congress Catalog Card Number: 99-72649

ISBN 1 85521 960 3

Printed in Great Britain

Contents

Understanding how to manage risk is crucial to survival and success in the modern world. ... Managing risk is a necessary new science - and it is not just for the technologists[1]

Preface

After the occurrence of great technical catastrophes - the Chernobyl reactor core melt-down, the Seveso poisoning, the Alaska coast pollution by a stranded oil tanker, the disastrous derailment of a high-speed railway train near Eschede in Lower Saxony are examples - investigations are usually started into the potential causes. Why did it happen, and how? Could it have been prevented? Frequently, human failure is identified as one of the causes. Somebody has made a mistake, somebody did not follow the rules, somebody did not inspect carefully enough the technical readiness of the equipment used.

In some cases, however, nobody is considered guilty of neglect. The disaster is said to have been caused by unforeseeable circumstances, by unpredictable events which, on the basis of the available knowledge, were completely unexpected. Nobody can be blamed, since it was unknown that the security precautions taken would turn out to be insufficient in the case under consideration.

We are all familiar with cases of both these types. They have in common that they occurred because they were connected with risks the magnitude of which had been wrongly assessed, if the risk had been seen at all. This could be due to negligence after all, or due to avoidable or unavoidable ignorance.

The analysis of unknown facts and circumstances, the search for sources of error, the estimate of limits of error within which a particular outcome may be expected to occur with a certain probability, is a task of science, as is the elaboration of proposals for the removal of sources of error and for the reduction of risks. Combatting ignorance is the task absolute of science.

Psychological aspects are of great importance in the prevention of catastrophes by human failure. As a consequence, psychology must be among the disciplines enlisted for an interdisciplinary investigation of various kinds of man-made or technology-made disasters, their origins, and their prevention.

---

[1] Benton, Peter, Riding the Whirlwind. Benton on Managing Turbulence. Oxford, 1990, p. 166.

Our time is characterized by an unprecedented progress of science and technology. The applications of scientific knowledge provide a high standard of living to a minority of the population of this earth, mostly in the Northern regions of the globe. The way of life of this minority has become the goal of development of the majority of people in the world who consider themselves underprivileged as long as the economic gap between the industrialized and the "developing countries" remains vast and is even increasing. But the life-style of the North, with its high consumption of non-renewable resources and its wasteful use of energy, is already causing serious damage to the natural environment. It cannot be a model for the growing world population because the balanced system of life on this planet would collapse if additional billions of people would adopt the life-style of North America or even Europe. Climate changes, destruction of the ozone layer, extinction of species of animals and plants, floods, droughts and hunger are forebodings of future catastrophes that are likely to happen unless the course of events is altered drastically. But how is this going to be accomplished? There are certainly a number of different options for the kind, and sequence, of policy changes and measures to be taken. In order to avoid unpleasant surprises, a thorough and unbiassed scientific analysis of these options will be required.

The situation is further confounded by the inherited nature of human beings. They are willing to co-operate with those whom they recognize as members of their own tribe or group. With them they are prepared to work together for common goals and joint enterprises. But they are equally inclined to fight bitterly against those whom they consider to be members of an opposing group, competitors for food, land, or work, or just suspicious "aliens" who do not share one's own beliefs, customs, habits, or values. They are declared to be enemies. Such "enemy images" are often upheld for centuries. Interestingly enough, however, peoples who have fought each other for centuries, like France and Germany, can become friends when a common enemy appears who seems to threaten all of them. Then they can be willing to join forces in order to meet the new challenge.

These characteristics were obviously essential for the survival of the human race in the course of evolution. Until very recently (on the time-scale of evolution) the number of humans on earth was very small, and so was the power of their tools and weapons. Apart from occasional fires, they could not do much damage to nature. Nature seemed overwhelming and inexhaustible. Furthermore, human beings were quite unable to endanger, by their own actions, the survival of the human race.

All this changed when, after the industrial revolution, modern technology supplied weapons of mass destruction, particularly nuclear, biological and chemical weapons, together with means of world-wide transportation and communication, at a time when the growth rate of world population had reached disquieting proportions. Finally, man had reached the ability not only to destroy the balance which nature had established in millions of years but also to commit homicide on a scale unthinkable in the past. It is highly disturbing that under these circumstances the traditional human tendency of maintaining "enemy images" and fighting against "hostile groups" continues. Under modern conditions this results in large-scale massacres, famines, and migration movements of refugees. By means of television reports, attempted relief operations, and by the arrival of masses of asylum seekers these events also destabilize world areas not directly involved in the original conflicts.

To turn around this dangerous situation for humankind, there seems to be only *o n e* hope and chance: The inhabitants of this planet will have to learn that they all belong to *o n e* group which has to keep peace and co-operate because it is threatened by a common enemy. This time the enemy is not coming from outside but from inside. It is the misuse, or the thoughtless use, of science and technology, irrespective of the undesirable by- and after-effects of often well-intended applications of science and technology. It is also the submittance to archaic instincts against people who do not seem to share our values although these people are our co-passengers in a boat struggling to stay afloat in stormy weather and needing all hands for bailing out water.

As a consequence of the situation described, humankind is faced with a twofold task of stupendous complexity:

1. to analyze the options available for action, particularly in economic and environmental policy, estimating for each option the risks, costs and benefits, taking into account by- and after-effects not only in the field of action chosen originally, but in neighbouring fields as well,
2. to study the history and psychology underlying the conflicts between neighbouring nations, and between ethnic and religious groups within nations, in time before war, or civil war, erupts, and to suggest joint tasks to be undertaken by the potential "enemies" which could make them understand that they are passengers in the same boat who depend on each other.

It is clear that the institutions of science (including psychology, anthropology and political science) and of letters (including history and economy) have an interdisciplinary and international mission here. Without their assistance, the twofold task cannot be completed, and catastrophes cannot be prevented. The political decision-makers by themselves cannot be expected to deal satisfactorily with the task. They cannot afford long-range thinking because they have to concentrate on winning the next elections. Long-range thinking often means accepting short-range sacrifices. This is unpopular as long as the electorate has not learned to honour such considerations. To explain to the public scientific, economic, historical and psychological facts, their likely consequences, the available options for dealing with them, and the obstacles that would have to be overcome, is a necessary task for which the academies of sciences and letters, and equivalent national societies and institutions, are better equipped than anybody else, particularly if they would organize interdisciplinary and international working groups for this purpose.

The author has dealt with the questions of scientific advice to decision-makers on global problems for a considerable number of years. In a series of lectures and publications, he advocated a stronger involvement of the international community of scientific and scholarly institutions in this field. Some of these observations are reproduced in this book. The first chapter, *Catastrophes as By-Products of Human Activities. How Can They Be Avoided?*, can be read as a general introduction. The other articles deal with different aspects of the twofold task described above, and of the responsibility of scientific and scholarly institutions for shouldering this task. All of them were originally written as papers and lectures for various conferences and workshops with different audiences. Some repetitions have been removed but, in the interest of smooth reading, not all of them. A certain overlap was unavoidable. In different contexts, the reader may come across the same or similar arguments. This may be taken as reinforcement and additional emphasis from different directions for the basic message of this book, and not necessarily as dispensable repetitions. Chapter 22, *The New World Order and the Role of Science*, summarizes this message and provides an outlook to the future.

Some of the chapters of this book are based on lectures given before the end of the Cold War. At that time the confrontation between NATO and the Warsaw Pact, between the USSR and the United States, was an obvious example for the confrontation between political powers which impute to each other threatening intentions incompatible with one's own peaceful and fully justified long-range goals, believing that they have to deter each other by military superiority. Today, one might have chosen examples from the armed conflicts in former Yugoslavia. The Soviet Union and the Warsaw Pact no longer exist. Therefore, it might seem that some of the papers written at a time when the nuclear superpowers confronted each other, are now obsolete. They were presented at international meetings and may have contributed in a small way towards détente and towards the final cessation of the Cold War, by showing that a stable peace cannot be achieved as long as threat perceptions are effective on both sides. Threat perceptions, it was shown, can only be gradually reduced by political means, not by armaments.

This, however, is a general fact, valid for all confrontations between armed powers. It is not limited to the special example of NATO-Warsaw Pact confrontation which, however, led to this conclusion in many East-West conferences of the past years. We decided, therefore, that it is appropriate to include these considerations from the recent history of the East-West conflict in this book which is devoted quite generally to the resolution of political conflicts and the prevention of catastrophes. There are lessons to be learned from the Cold War. They should be taken into account, e.g., in the present situation after the Kosovo War. The Cold War ended when both sides realized that their opponent had no real interest in aggression.

Thus, the East-West confrontation of the second half of the 20th century remains a good example of conflicts based on mutual suspicions of the aggressive long-range political goals of the respective adversaries. The discussions of that time, referred to in this book, continue to be an example of a scientific analysis of a dangerous conflict situation. They were held in order to identify options for the prevention of escalation and of the disastrous use of military force. The conclusions drawn, e.g., in chapter 12 of this book, *Changing of Long-Range Political Goals as a Prerequisite for Progress in Arms Control*, remain valid for all conflicts of this type. A particularly good example of the recommended scientific method for the identification of the different assumptions and perceptions underlying differences of political opinions, and of judgements regarding the feasibility or advisability of certain political measures, is given by our Workshop on President Reagan's "Strategic Defense Initiative (SDI)", referred to in chapter 13 of this book, *SDI and Stability. The Role of Assumptions and Perceptions*.

To facilitate the use of the present publication as a handbook of threats and dangers, of potential methods to avoid them, and of possible ways in which the institutions of science could contribute towards this goal, "Topical Summaries" are added as chapter 23. This makes it easy for the reader to find within the remaining 22 chapters those places in which he or she is interested, i.e. where:

(a) questions which are important in the context of avoiding catastrophes and conflicts and which need further research,

(b) known facts regarding the psychological prerequisites for keeping peace between competing groups,

(c) the dangers threatening humankind at the beginning of the 21st century, including threats to the natural environment, and

(d) the resulting tasks for the institutions of science and letters

are dealt with. For better orientation, brief quotations of relevant parts of the statements in these places are given.

It is hoped that with these aids this book will serve its purpose of helping to avoid future catastrophes and conflicts, and of stimulating further the international community of scientific institutions to use to the fullest extent possible its resources of knowledge and logistics in this essential field of service to humanity.

I am much indebted to the publishing houses *Campus Verlag (Mr. Thomas C. Schwoerer), Nomos Verlag (Dr. Volker Schwärz), Akadémiai Kiadó Rt. (Ms. Piroska Polyánszky), Indicator Press (Mr. Ted Leggett) and Plenum Publishing Corp. (Prof. Herbert H. Krauss and Ms. Georgia Prince)* who allowed the reproduction of articles previously published by them,[2] and to the *Pugwash Conferences on Science and World Affairs* as well as the *International Amaldi Conferences of National Academies of Sciences and National Scientific Societies* which quite generally allow the reproduction of papers from their proceedings, asking only for the acknowledgement of the original source.

Klaus Gottstein   *August 1999*

---

[2] Footnotes at the beginning of each chapter indicate where the respective article was first published.

1 Catastrophes as By-Products of Human Activities

How can they be avoided? [3]

For thousands of years human beings and their settlements have been threatened by natural catastrophes. Many examples come to mind: the eruption of the volcano of Santorin (Thera) around 1500 B.C. which caused a huge wave by which very probably the Palace of Knossos and the Minoan Civilization were destroyed (origin of the saga of Atlantis?), or the end of Pompei and Herculaneum by the eruption of Mount Vesuvius in A.D. 79, and in our days earthquakes and floods, typhoons and hurricanes of all kinds. But also man himself, already thousands of years ago, has brought about large-scale environmental devastations, for instance by causing forest fires. Up to the industrial revolution, however, the technical abilities of man were limited, so that the impact of his activities on the biosphere of the Earth was negligible. Human and animal living conditions were not much impaired by them. In the meantime the situation has been changed drastically by the development of human technologies and by population growth. In principle, humankind is capable today to cut down all rain forests and thereby change the climate of the earth, to destroy the ozone layer, to contaminate radioactively the surface of the earth and thereby jeopardize its own survival. Humans did not cause these perils consciously and intentionally or light-heartedly, but only by using, as all earlier generations did, their capabilities for exploring their environment and improving their living conditions. In earlier times, however, when their present surroundings were sufficiently exploited, they could move on into areas still untouched. This possibility no longer exists. Nature can no longer be considered inexhaustible, as it seemed to be in the early history of mankind. Then, as long as nature was all-powerful and the number of humans relatively small and poorly equipped, man could exploit nature without worrying about after-effects. Joseph Alois Schumpeter (1883-1950), one of the best-known economists of the 20th century, still held that "nature and the families do not write bills, one may

---

[3] Lecture given in German at the meeting "Environmental Destruction and Proneness to Catastrophes" of the German Committee for the International Decade for Natural Disaster Relief (IDNDR) of the United Nations, Haus Bommerholz, Witten, Germany, 11-12 November 1991. (Translated and slightly extended by the author)

exploit them at no cost". Today, this is no longer true. Nature must even be artificially maintained and kept alive by conscious policies. Under the rule of man, it no longer maintains itself. If man continues single-mindedly to pursue his goals, as he has been used to doing since primeval times, without thinking of the by- and after-effects of his activities, then these by- and after-effects will lead to catastrophes which will end his career on this planet. Already today the great problems of humankind may be seen as the unintended consequences of human activities: in a certain sense our environmental problems are the inadvertent by-effects of our aspirations to obtain more food and more comforts, overpopulation is a by-effect of our fight against diseases and for better sanitation, and the disquieting arms race a result of the quest for more security. On the other hand science and technology gave to humankind the chances for a better quality and a longer duration of life, chances of which earlier generations could only dream. Moreover, in many cases science and technology allow the easing of consequences of natural catastrophes by predicting, for instance, the occurrence of hurricanes and floods and sometimes even the eruption of volcanoes, and thereby making early evacuations possible.

With the assistance of science and technology it would no doubt be feasible in many cases also to foresee the occurrence of by- and after-effects of human activities. They could be taken into account and made part of the overall planning. Countermeasures could be prepared, or the activities planned could be replaced by others with less harmful consequences. As the main effects and the by-effects often occur in different fields of specialization (in security policy and in psychology, for instance, or in economy and in climate research), multidimensional thinking and interdisciplinary collaboration are required. Because of the global character of many problems, international co-operation is also indicated. In economic policy it is also necessary because laws in the field of environmental protection must be valid worldwide if they are to be effective. Environmental protection costs money. If a country would not enact laws which are valid in other countries, its industries would have an economic advantage. In order to maintain its economic competitiveness, no country would be prepared to set up effective environmental protection unless its competitors would be subject to similar laws.

For species of animals and plants threatened by extinction sufficiently large protection areas and natural habitats have to be created by international agreements. If this causes disadvantages for the people living there (because they are prevented, for instance, from developing their neighbourhood), financial compensations or new homes may have to be supplied. Since every place on earth has become accessible to man and his nature-conquering machines, man himself must impose on himself certain self-restrictions in order to protect the biosphere. A significant rest of original nature should be preserved. Nature has been the living space of human beings from time immemorial. Here they developed their characteristic physical and mental properties. In order to be able to understand and control these properties, the sight of undisturbed nature should remain visible in some places.

The costs of ecological by- and after-effects should be taken into account by industry and commerce when planning new investments. Consideration for ecology should be inseparable from a socially minded market economy at the end of the 20th century. The framework within which this will be possible must be set, on an international level, by the community of states. Each state will have to set positive tasks of environmental protection which can be solved within the market economy.

International co-operation, if it is to be successful, requires psychologically trained understanding of cultural differences and of differences in perception. This is particularly true for the estimation of risks from so-called human failures in the operation of large-scale technologies such as nuclear power stations, oil supertankers, arsenals of ABC weapons. In cases of this type it is not sufficient to set arbitrary safety factors in order to minimize the risk of accidents. It will rather be necessary to investigate all possibilities for errors in the behaviour of all members of all potential operation-, maintenance- and repair-teams, taking into account their cultural backgrounds and psychological attitudes. Even after appropriate measures were taken it will still be necessary to include a still thinkable "worst possible out-turn"[4] into the analysis. "Managing risk" is a new science which will be of increasing importance in years to come.

It cannot be the task of science, of course, to decide which risks society ought to take in order to reap which benefits, or, in other words, which costs should be considered tolerable for gains of some sorts. This is a decision based on values which can only be taken by the public, or by the politicians representing the public, after understanding the results of a thorough and carefully explained cost-benefit analysis. Science cannot relieve politics from the responsibility for these value-based decisions. What science can (and should) do, however, is to describe the available options, and to submit each option to a cost-benefit analysis, so that the consequences and by-effects to be expected become clearly visible.

---

[4] Benton, Peter, Riding the Whirlwind, Oxford 1990, page 158.

So far the institutions of science are inadequately prepared for this mission. Nevertheless, there are some hopeful beginnings. The National Academy of Sciences of the United States, at the request of the U.S. Government or on its own initiative, has, for many years, taken up questions for the solution of which scientific advice is obviously required, and set up committees for their analysis. It maintains, as permanent institutions, a Committee on International Security and Arms Control (CISAC) and a Committee on Science, Engineering and Public Policy (COSEPUP). For many years CISAC has met regularly, about twice a year, with a similar committee of the Academy of Sciences of the USSR now of Russia. In these meetings technical problems of arms control, of verification of disarmament measures, of registration of nuclear test explosions, etc., were discussed. The results are put at the disposal of governments. Certainly these communications had an important role to play in the preparations for the official negotiations between the two governments.

Some European academies and national scientific societies have found themselves ready in recent years to take part in such discussions of experts on scientific questions of political relevance. They followed an initiative taken by the President of the U.S. National Academy of Sciences, Dr. Frank Press who, in 1986, invited the European academies and scientific societies to form a European committee similar to CISAC and take part in the U.S.-Soviet discussions on arms control and disarmament. Although an agreement on the creation of a joint committee of the European academies of sciences on questions of arms control could not be reached, the Accademia Nazionale dei Lincei in Rome under its Vice-President, later President, Edoardo Amaldi, formed an Italian committee on international security and arms control and invited the national academies and scientific societies to international conferences which took place in Rome in 1988, 1989, and 1990. For the first of these conferences only institutions of Western Europe and the U.S. had been invited, in the later conferences also representatives of the academies of sciences of Eastern Europe, of China, and of the Third World Academy of Sciences in Trieste took part. Questions of arms control were the main subjects of these conferences, but scientific-technological questions of environmental protection and of co-operation between industrial and developing countries were also beginning to appear on the agenda. Professor Amaldi died while preparing the third of these conferences (Rome, 1990). His successor as President of the Accademia Nazionale dei Lincei, Professor Giorgio Salvini, continued the preparations, and the guardianship for the following conferences of this type which, in honour of their founder, were called Amaldi Conferences from then on.[5] After initial hesitations, scientific institutions of other nations agreed to follow the example of their Italian sister and to host Amaldi Conferences.

It is the proposal of the present author to extend the agenda of Amaldi Conferences, after careful preparations by international and inter-disciplinary working groups, and in addition to the pressing problems of arms control, to the scientific aspects of other problems of global significance and great urgency, or to organize similar conferences of the international community of national academies of sciences and national scientific societies, possibly under a different name, for this purpose. The topics to be analyzed and discussed could include the following:

- Risk assessment, risk-benefit analysis and discussion of possibilities for international sharing of great risks (oil supertankers, nuclear power stations, climatic changes by the destruction of rain forests, etc.) In many cases, risks of this type could be strongly reduced by expensive precautionary measures. The obligatory introduction of such measures would require international agreements on rules and regulations for the reduction of risks, and for the distribution of the financial burden involved,
- Determination of the size and quality of reservations and natural habitats necessary for the preservation of animals and plants threatened by extinction,
- Rough determination of the minimum volume of biomass (trees and plants) required for the preservation of a balanced relation of the combinations of oxygen, carbon and nitrogen in the atmosphere and for the prevention of undesirable changes of climate,
- Drafting of a list of geographical regions which, according to the criteria mentioned above, would have to be exempted from industrialization and dense population,
- Juridical and economic investigations about possibilities for compensating inhabitants of regions which, because of the protective measures indicated above would be denied development and industrialization, and might even have to be resettled,
- Investigations by experts of international law on possibilities for the protection of such reservations from poachers and other intruders (Green Helmets of the United Nations?),

---

[5] Venues of later Amaldi Conferences: Cambridge, England (1991), Heidelberg (1992), Rome (1993), Warsaw (1994), Piacenza (1995), Geneva (1996), Paris (1997), Moscow (1998).

- Questions of the conversion of arms factories to civilian production,
- Economic problems of East-West co-operation,
- Special topics of scientific-technological co-operation between industrialized and developing countries.

Under present conditions a programme of this scope surpasses by far the capabilities of academies and national scientific societies. On the other hand, the problems exist! They cannot be solved satisfactorily without the scientific expertise available to academies and scientific societies. In the long run, therefore, the latter will not be able to evade the responsibility to share their potential knowledge and their methods with those who, faced with impending catastrophes, need them for making the right decisions. In order to avoid confusion it may be advisable to start step by step. But time is running short. To begin with, a precise definition of the problems to be analysed and a stock-taking of the expertise and the capabilities internationally available for their treatment will be required. The procedure for the selection of the problems to be dealt with in international and interdisciplinary scientific co-operation should be established by international agreement between the academies and national scientific societies.

Summarizing, one might state the following:

Population growth, weapons of mass destruction, environmental destruction, and social tensions between the rich and the poor in developing countries, and between the affluent countries of the North and the majority of humankind living in the hungry Third World, have contributed to a situation in which the total system representing the human race is greatly endangered. Modern methods of transportation and communication see to it that catastrophes in one part of the world do no longer, as in earlier times, leave the other parts undisturbed. Today, television reports, economic repercussions and masses of arriving refugees make them felt, in other regions as well, in a destabilizing way.

Essential prerequisite for avoiding, or at least limiting, the perils associated with a collapse of parts of the complex network of modern society, is preservation of the knowledge necessary for operating all components of the network. Not only technical know-how is required but also managerial, organisational, and psychological knowledge. Of great importance is the rooting of an ethical basis in a population from which the persons to be responsible for the control, further development and improvement of the network must be recruited. Accuracy, reliability, sense of duty are indispensable qualities of the "service personnel" for the machinery of modern society which is so dependent on applications of science and technology. The cultivation of ethics and knowledge requires stable institutions of cultural, scientific and technical education. Without them, attitudes and knowledge which is required for the avoidance of catastrophes and which can only be transferred from one generation to the next by cultural means, might get lost. Only animals inherit biologically the information necessary for their survival in their natural habitat. Humans have to learn from their teachers. Securing the stability of educational institutions is therefore of paramount importance in our technical age. The stability of the institutions of society in general is conducive for a considerate, circumspective decision-making under the difficult conditions of an overpopulated world with shrinking distances, endangered by disasters of different kinds, in which peaceful co-existence will only be possible on a foundation of tolerance, consideration, empathy, understanding and readiness for sacrifices.

Stability does not mean stagnation. Institutions, attitudes and the rules of the game have to be adapted to changing conditions. Otherwise, eruptions will lead to just those destructions which are to be avoided in the interest of an orderly transfer of knowledge and ethical attitudes. Reforms will always be necessary. But they should be carried out step by step - as were the reforms in 19th century England - so that the consequences remain within sight and corrections are feasible when unexpected by- and after-effects occur. It is always hazardous in politics to follow utopias, i.e., to have definite goals of desirable futures which must be reached recklessly whatever the cost.[6] It means putting all eggs in one basket and being unprepared when reality destroys utopia. This can have disastrous consequences. Deep skepticism is indicated whenever utopian goals are taken serious.

In our scientific-technological age human actions are no longer sufficiently restrained by boundaries set by nature. If catastrophes are to be avoided, artificial safeguards are required. The best safeguard is strict adherence to the principles of democracy: Majority rule, protection of minorities, free competition of ideas, legal opposition. Where the conditions of a working democracy do not yet exist, they have to be created step by step. The artefact of political culture offers to humankind the only chance of survival.[7]

---

[6] Christian Graf von Krockow, Politik und menschliche Natur, Stuttgart 1987.
[7] Christian Graf von Krockow, *loc.cit.*

2 Pluralism and Pugnacity [8]

The collapse of central authority,
what it entails,
and the task of scientific institutions

Anthropology and history show that independent groups of people, when confronted with problems of a similar nature, do not always arrive at similar solutions. With the same brain capacity, and often even with the same environmental background, they developed different languages, philosophies, religions, social structures, and political theories.

That independent thinking can lead to highly differing results was already discovered in antiquity. Antisthenes and Aristippus, both of them friends and pupils of Socrates and deeply influenced by his teachings, founded schools, after the death of Socrates, with diametrically opposed recommendations as to the best way to conduct life: The Cynics of Antisthenes believed in the supreme value of virtue, poverty and hard word. The Cyrenaics of Aristippus taught the hedonistic enjoyment of the pleasures of life, as their master, not slave. Similarly, the school of the Epicureans and the Stoics, which flourished simultaneously, offered opposite recipes for happiness to their Greek contemporaries who were looking for guidance in a period of transition and change of values. Whereas the Epicureans urged the withdrawal from public life to find quiet of mind in sensations that produced pleasure without pain or annoyance, the Stoics taught insensitivity to pain and devotion to duty and to engagement in public affairs. The rise of quite different philosophies from the same situation was not limited to antiquity. In the 19th century, marxists, liberals and conservatives competed with differing theories about the best way to run the economy and the society. Our century saw the rise and decline of different political theories, ideologies and world views. At present, we witness all over the world, with a few exceptions, declarations of faith in democracy and market economy. These systems are supposed to satisfy the demands of the people better than all other known systems, particularly when they are linked with social

---

[8] Paper presented at the 42nd Pugwash conference on Science and World Affairs, Berlin, Germany, 11-17 September 1992. Published in J. Rotblat (Editor), Shaping our Common Future: Dangers and Opportunities. World Scientific Publishing Co, Singapore London 1994.

15

legislation. But we also witness the re-emergence of nationalistic strife which had long been suppressed and assumed forgotten.

The variety of answers given by human thinkers to the questions on nature, God, society and human existence led the ancient philosopher Pyrrho and his school of Skeptics to the conclusion that truth cannot be recognized so that one should abstain from judgement and be prepared to doubt any perception. According to him, dogmatists should be healed from their rashness and imaginations.

In our days, such skeptic moderation is rare. In general, the multiplicity of views, interests and traditions results in strife when groups representing them get in touch with each other. Strife and fighting can be avoided, however, if the opponents have come to belong to a social system with a strong central authority that sets rules for a peaceful settlement of conflicts, and establishes enforceable sanctions against violations of the rules. Central authorities of this type exercise a monopoly of power within a certain region and thereby prevent armed conflicts among the tribes, territories or member states under their jurisdiction. In history, these authorities have sometimes been based on military force and dictatorship, as in the cases of the Roman Empire before its decline, and the Soviet bloc under Stalin. Sometimes, the authority derived from voluntary co-operation under a mutually accepted constitution or negotiated agreements, as in the cases of the United States of America and the European Communities. In our days, we witness again, this time in the former Soviet Union and in the former Yugoslavia, the consequences of an unpremeditated and unprepared dissolution of a previously reasonably efficient, though economically ineffective and widely unpopular, central authority.

It may be of interest, in this context, to have a look at the way in which the unification of the former two German states was performed: The dissolution of East Germany as an independent state and its merger with West Germany was carried out on the basis of international agreements and with the consent of the neighbours and allies of both German states, and with the strong support of the majority of the East German population. Moreover, political, administrative, and financial details of the merger were fixed beforehand in extended negotiations between representatives of the East German and West German governments and by legislation of both parliaments. Although the re-unification after 45 years of separation under two entirely different social and economic systems met, and still meets, with many practical difficulties, injustices and reasons for dissatisfaction, it was accomplished without armed insurrections and without any bloodshed. Certainly, the conditions extant in Germany at that time were special and cannot be reproduced in other parts of the world. But the lesson should be learned from developments in the former Soviet Union and in Yugoslavia that it is extremely dangerous to dissolve an established order before the new order that is to follow is agreed upon by the parties affected. No effort should be spared to create an acceptable framework for a new order in cases where strong, quarrelling factions demand the dissolution of a political organism no longer viable. The leaders of such factions should be made aware of the dire consequences in potential bloodshed and widespread destruction that will follow if the old order is destroyed before a new one is accepted to which an orderly transition becomes possible.

It must be admitted that Gorbachev tried to build, or keep, a central framework within which the republics constituting the former Soviet Union could have developed their national identities in an orderly, co-operative, and mutually advantageous fashion. But Gorbachev failed. The republics became sovereign, and there is no guarantee that they will find it to their advantage, in the long run, to co-operate peacefully. Warfare with many victims between Armenia and Azerbaijan, civil war in Georgia, disputes between Russia and the Ukraine regarding possession of the Crimea and the Black Sea Fleet give rise to deep concern. Historical experience shows that nations as well as politicians are ready to resort to arms when they believe that their interests are at stake and when one of the opponents feels that armed fighting will give him a chance to win or to avert defeat. This has not changed in our days. More than 100 wars and major armed conflicts have taken place since the end of World War II. Since ancient times up to our century poets, philosophers and peace movements of ordinary people have deplored the devastations of war, without much avail. Wars continued, and some other poets, philosophers and ordinary people even glorified war as the great innovator, the father of all things (Heraclitus) and the instrument of social Darwinism. Under these circumstances, it is not surprising that warlike conflicts continue in several parts of the world, and now also affect Europe. It is rather the extraordinary fact that Europe was spared war for 45 years after 1946 that will have to be explained by future historians. Peace was maintained in Europe although on its soil two heavily armed blocs confronted each other, each led by a nuclear superpower. Peace was maintained in spite of ideologies on both sides that declared the system of the other side, respectively, as doomed and ready to be toppled in order to find its proper place on the garbage heap of history.

It seems that mutual deterrence by nuclear weapons, i.e. the spectre of joint suicide in the case of armed conflict, made a major contribution towards the wise restraint exercised by the leaderships of both sides in dealing with each other during the period of Cold War. Of course, there were other good reasons for not waging war. But similar good reasons for not waging war had existed prior to many wars which, nevertheless, broke out. If nuclear deterrence was instrumental for avoiding war between NATO and the Warsaw Pact, it was indeed an extremely risky method. Had deterrence failed, the use of the existing nuclear weapons could have wiped out human civilization if not the human race.

A precondition for keeping nuclear deterrence under control was the existence, on both sides, of a stable political leadership giving directions to a well-disciplined military establishment responsible for safeguarding the nuclear warheads and carriers.

After the dissolution of the Soviet Union, the stability and the authority of the governments of the new republics as well as the discipline and obedience of the military forces have become questionable. But the nuclear weapons of the former Soviet Union are still around. There is a finite risk that their existence may be used, not for deterring aggression and in order to maintain a stable equilibrium as between the superpowers during the Cold War, but as a means of blackmail against weaker neighbours. Under such circumstances and due to a failure in the chain of communication and command, or by irresponsible action of some commanding officers, nuclear weapons may actually be detonated, with unthinkable consequences.

It must be of paramount importance, therefore, to create a political atmosphere in which governments and populations feel secure, in which no threats are perceived that would seem to justify military countermeasures. In those cases in which old animosities exist, every effort should be made to establish some sort of truce, as formerly between personal foes in a beleaguered castle. In our time, the role of the common enemy necessitating that truce may be assigned to the joint problem of managing survival in the presence of economic and ecological threats of regional and global scopes, cultural clashes, unequal distribution of resources etc. These challenges can only be met adequately by sober, honest and peaceful co-operation. Western Europe, after the Second World War, reached reconciliation and unification under the perception of a common threat emanating from the Soviet Union. In the Soviet Union and in Eastern Europe, fear was fomented that "the imperialists" might launch another aggressive war. This expectation, and the efforts to foil the imperialistic plans, seemed to justify the acceptance of economic and social hardships, the limitation of human rights, and the subordination of national interests to those of the "socialist camp". The solidarity of the "socialist camp" collapsed when it was no longer enforceable. It was no longer enforceable when it became evident, after decades of peace preservation, that the aggressive plans of "the imperialists" did not really exist. This means that there was no longer a common enemy against whose evil designs a military pact of unequal partners was necessary and justifiable. It is not clear yet if, and how, in the long run, NATO will survive the demise of its traditional opponent.

In other words, the nations of both Eastern and Western Europe, and as a matter of fact those of all other parts of the world, require a new sense of partnership. They all face the huge challenges of our time which can only be mastered by international and interdisciplinary co-operation. Hunger in the Third World, deforestation, destruction of the ozone layer, risks from inadequately designed and maintained nuclear reactors, the rapid progress in the extinction of species, illiteracy in the presence of high technology, drugs and diseases are only a random selection of those problems that may be considered common enemies of humankind. They should replace, in human consciousness, the more traditional enemies to whom each group of people, tribe or nation had become used in the course of centuries. Only when this replacement will have succeeded, joint action for tackling the great problems of our time will become possible. This, however, will require joint institutions for mediation and judgement. As shown above, truly independent and sovereign institutions will develop different ideas and conflicting interests. There must be a higher level on which such conflicts can be resolved. This may be a higher administrative unit entitled to make enforceable decisions about the dissenting opinions of sub-units. It may also be a parliamentary assembly of a superordinated unit or an International Supreme Court whose rulings are binding and enforceable by an international police force. This may still sound utopian, but in the presence of the conflicting interests and of the problems of our time, there is no alternative but war and destruction to the creation of effective international institutions. They may operate by the substitution principle, i.e. leaving all decisions to national, regional and local levels that can be dealt with effectively on those levels, and assuming jurisdiction only on those issues that need decision-making on the highest level because conflicting interest at the lower levels prevent measures which are considered necessary in the interest of the higher unit.

The decision-making and judgement-passing authorities at the higher levels, that are supposed to stand above the conflicting interests of the lower levels, will require guidance as to which solutions are appropriate. Parliamentarians at the higher levels who will have to pass legislation to be considered by the Courts of Justice of their level will also need expert advice. This is a challenge for scientific institutions to form international interdisciplinary committees of scientists, engineers and scholars (including, if necessary, economists, psychologists, historians, political scientists, international lawyers) who devote themselves in an impartial way to the task of developing options for political action in the cases under consideration. For each option, the costs, risks and benefits should be stated, taking into account, if possible, short-term and long-term side- and after-effects. This should include investigations of the potential long-term effects of measures in one field on the conditions in another field. For example, the potential impact of technical projects on the natural and social environments, or the potential consequences of economic decisions for social stability could be studied in depth. Of course, as Pyrrho and the Skeptics had surmised, though for other reasons, there are questions to which the answer cannot possibly be known, because too many variables are involved, or because Heisenberg's uncertainty principle has to be taken into account. The future fate of humankind is among the things we cannot know in any detail for this reason. There are, however, potential developments that can be foretold because they are the consequence of human actions. The stress here is on the word "potential" because also in these cases it is often impossible to predict what will happen when human actions are involved. But it is often possible to anticipate what *might* happen within a certain range of probabilities. If a major catastrophe is among these possibilities with a non-negligible probability, then it would be irresponsible to assume that this event will not happen. In such cases, the urgent scientific advice should be given to take a less risky course, and the options available for reducing the risk should be described.

It is true that there is already much talk about a "technology assessment" of this kind. But, so far, little has been achieved in practice, and the existing scientific institutions, with some exceptions, have not yet found the proper channels to make the know-how, which is at their disposal, accessible to the decision-making process. There are some hopeful signs that this situation is now recognized, but much is left to be done.

3 New Patterns of Conflict, Old Methods of Conflict Resolution? [9]

The necessity of new approaches based on a more thorough use of the resources of natural and political science, psychology, history and international law

The end of the Cold War and of hostile confrontation of the nuclear superpowers has had spectacular results. Among them was the dissolution of the Soviet Union of Yugoslavia and of Czechoslovakia and the unification of Germany. It is too early to say whether there will be more territorial changes in the wake of this historical turnover.

The separation of Czechoslovakia into two independent states and the merger of the former two German states were carefully prepared by parliamentary procedures and international agreements and were carried out without bloodshed. On the other hand, the dissolution of the Soviet Union and of Yugoslavia resulted in local wars in the Caucasus region, in bloody persecution of minorities in some of the Asiatic republics of the former Soviet Union, and in war between Serbs, Croats and Muslims in former Yugoslavia. Nobody knows whether these conflicts will spread, and whether more warfare will follow. Efforts by the United Nations and by the European Community to stop the fighting so far remained without success. Is a new pattern of conflict visible here? Or are we witnessing just the old type of conflict which history has seen over and over again, the struggle for power and for survival between different human groupings kindled by reckless politicians for their own ambitious designs? Have we returned to a period in which the "Law of the Stronger Power" is the supreme law? Will future conflicts be solved by brute force? Will peace be dictated on the basis of

---

[9] Paper presented at the 43rd Pugwash Conference on Science and World Affairs, Hasseludden, Sweden, 9-15 June 1993. Published in J. Rotblat, Sven Hellman (Editors), A World at the Crossroads: New Conflicts, New Solutions. World Scientific Publishing Co., Singapore London 1994.

23

"ethnic cleansing"? Or will it be possible to end armed conflicts of the type now raging in Yugoslavia, in Georgia, Armenia and Azerbaijan, by mere negotiations?

During the Cold War both sides were interested in maintaining the status quo, and in avoiding Hot War which was seen as leading, inevitably, to nuclear holocaust. Thus, negotiations were the only available method for settling disputes, apart from propaganda campaigns, diversion, economic warfare and conventional wars by proxy in the Third World. These negotiations, particularly on questions of arms control and confidence-building measures, often lasted for many years and led to meagre results. Nevertheless, open war between the two blocs was avoided. Neither side had wanted war, considering the risk of self-annihilation. Under these circumstances negotiations made sense. Does this experience apply when fighting is already in progress, and when at least one of the fighting parties is not interested in peace, or even in an armistice, before final victory? It seems this is the situation in some areas of the former Soviet Union and of former Yugoslavia.

What does this mean? May we, nevertheless, leave these conflicts to the old methods of negotiation, knowing full well that they cannot end the fighting because the fighting parties do not trust each other? Obviously, at least one of the parties in each conflict believes that continuing the fight, not stopping it, is the only way to avoid annihilation. What is the origin of the armed conflicts which arose after the collapse of the former central authorities in Yugoslavia and in the former Soviet Union? What are the available options for preventing local wars of this type, and for stopping them once they have erupted?

The medley of peoples, cultures, races and religions in the Balkans is the result of more than thousand years of history abounding with wars and occupation by successive empires. The Byzantine empire, the Turks, the Habsburgs, among others, were involved and left their traces. The colourful multiplicity of peoples, languages and cultures in the former Soviet Union reflects several centuries of a policy of expansion, settlement and modernization by the Tsars who, in addition to gaining new subjects by annexation, called in settlers, artisans and scholars from the West, among them many Germans, who became loyal Russian citizens. In closed settlements they often preserved the language and culture of their original home country. In many parts of the world, wherever the whims of history resulted in the close proximity of groups representing different cultures, similar mixtures of populations can be found.

Such neighbours of different origin usually lived peacefully side by side, or even gradually intermingled, as long as the type of government continued under which they had become neighbours. Rulers like the Tsar at St. Petersburg, the Sultan at Istanbul and the Emperor in Vienna saw to it that law and order among their subjects were preserved - which did not exclude occasional uprisings and latent tensions. Under this protection, in any case, individual ethnic groups were able to preserve their identity over many generations.

New and often artificial borders had been drawn many times in history. But, at least in modern times, the population had been left in place after a change of sovereign. Only in our century was the method of "ethnic cleansing" introduced. It was first applied by Hitler, Mussolini and Stalin who, for political reasons, moved populations from the territories where they and their ancestors had lived for centuries. The ground for these arbitrary cruelties had been prepared by the rise of nationalism and racism in the 19th century. Earlier only "religious cleansing" had been known. The Huguenots, the Salzburg Protestants, the Puritans had been made to leave their home countries because of their religious faith which they were unwilling to change. Jews were able to escape persecution if they accepted Christian baptism. In the 19th century, however, the idea of racism, of superiority of one race over the other, was promulgated (J. A. Gobineau, Houston Stewart Chamberlain and others). Social Darwinism and National Socialism followed.

The idea of racial purity or cultural homogeneousness, on which the practice of forcible "ethnic cleansing" is based, must be considered absurd in our age of high mobility by modern means of transportation. In former times when even moderate distances were hard to overcome by ordinary people, tribes of even small size and their cultures could possibly remain unaffected by outside influences over long periods of time, and sometimes they could form separate nations. At the end of the 20th century the tendency is in the opposite direction: Even the old, great national powers of Western Europe - Great Britain, France and Spain - as well as Italy and Germany (united only in the second half of the 19th century) find it difficult today to preserve their national and cultural identities under the impact of massive foreign immigration. They are willing to part with important prerogatives of national sovereignty for the sake of promoting European unification. This willingness to move away from nationalism toward European integration is the fruit of historical experience: Fully independent states are inclined to follow their own interests at the expense of other states, and this leads to conflicts and, if the respective leaders feel strong or desperate enough, these conflicts lead to wars. This is what the newly independent states of former Yugoslavia and of the former Soviet Union are now experiencing, and this is why European history shows a long series of bloody wars between European national states, and sometimes even between smaller independent units (e.g. in Italy and Germany before these countries were united).

Creating new, fully independent nation states or, as was proposed for Bosnia-Hercegovina, autonomous provinces based on the results of "ethnic cleansing", is therefore likely to lead to a continuation of fighting. It is old thinking. Giving cultural self-determination to formerly suppressed ethnic groups, while protecting minorities within these groups, is timely and necessary. But if there is a risk that ethnic groups will take up arms against each other because of old enemy images they are projecting upon each other, then peace can only be secured if the following conditions are fulfilled:

1. The use of arms must remain the monopoly of an authority at a level higher than that at which the potential enemies are located. Otherwise the war, when started, will probably be fought to the bitter end. In the case of Yugoslavia, e.g., this would mean that the war-fighting parties would have to be disarmed and that the embattled areas would have to be policed by an international (United Nations or European) force until a reliable and enforceable peace settlement had been found. This international police force would have to be ready to use its arms for the protection of peace.

2. Parties to the conflict must be made to agree to settling their differences by bringing their cases to an International Court of Justice, and by recognizing as binding the judgements passed by this Court. Its decisions must be enforceable.

3. All possible steps must be taken to reduce existing enemy images which are the source of hatred, misunderstanding and fear.

History shows that the risk of war is often very real among neighbouring states. With the advent of modern weapons of mass destruction and with the complexity and vulnerability of modern society the traditional institution of war must be considered obsolete if human civilization is to be preserved. It must be the goal of international politics, therefore, to reach general acceptance and fulfillment of the three conditions mentioned above. This may seem difficult and even unrealistic at the present time. But there is no other way if continued fighting with its devastating effects - human suffering, psychological damage and destruction of cultural treasures - is to be avoided.

If one or both of the fighting parties in a local war do not accept conditions 1 and 2, then, as mentioned above, an international police force would have to be sent in to end, by applying force, the use of arms. The only alternative would be, in order at least to prevent the conflict from spreading, to isolate it and to cut off supplies from outside. This, however, has dire consequences on the morale of the general public all over the world. The daily show of cruelties in the mass media, the open violation of all rules of decency with no sanctions following, will have a devastating effect, particularly on young persons in our own countries. A few people may become motivated to join humanitarian movements but the majority, it seems, just becomes insensitive. Already there is an increasing number of reports on acts of violence in our streets. Victims of attacks do not get any assistance from bystanders who are so used to unpunished violence from watching the news on TV that they lose their moral standards in real life as well. They get used to the role as passive spectator to any act of gross violation of a traditional code of conduct which, step by step, becomes irrelevant. This is very dangerous since a society without moral standards will produce dictators, or will collapse.

Psychologists, historians and political and social scientists can help with the elimination of enemy images. They can explain the origin of threat perceptions and the way in which fears and misunderstandings reinforce each other and lead to preventive or retaliatory actions which are interpreted as aggressive by the other side and thus seem to confirm the original enemy image. They can also explain, to the general public, the role of enemy images for rallying the public behind a demagogic leader. Sometimes enemy images derive from a feeling of superiority which is cultivated by minority populations with respect to the majority around them and which serves to prevent intermingling and thus to preserve their cultural identity. This can be explained, made conscious and thereby deprived of its potential explosiveness.

Another important phenomenon known to psychologists but often ignored by administrators and politicians responsible for the accomodation of foreign refugees is the unconscious occurrence of fear and aggression in any community in which strangers suddenly appear in numbers exceeding 20 per cent of the total population. This may be observed in subway carriages as well as in Bavarian villages.

In addition to explaining, and putting into perspective, historical, psychological, geographical and other scientific facts underlying enemy images, scientists and scholars must call attention to the global and universal problems of environmental pollution, hunger, disease, drugs, population growth, energy shortage and related problems which really need every conceivable effort. The group feeling of solidarity created by enemy images should rather be directed towards these threatening global problems. Scientists and scholars have an important role here, not only in describing these problems but in designing strategies for dealing with them. These strategies should be based on interdisciplinary studies by local, national and international experts. They should assess by-effects and after-effects of measures considered for implementation, as well as the risks and benefits involved. This should result in options for action. Co-operation of different factions of the population in the discussion of these studies and in planning the implementation of measures could create feelings of partnership and assist in the elimination of enemy images.

The highest priority in efforts of this kind should be given to the building and maintenance of a socially stable society. It is only in a stable society that the environment can be protected. Moreover, modern civilization requires the availability of well-trained experts for the maintenance of power stations, chemical plants, railway and airline services, oil pipelines and tankers and other facilities. Only a stable society can provide the high standard of education needed for the supply of well-trained experts. Without them, environmental catastrophes, traffic disasters, electricity blackouts and other malfunctions will lead to a collapse of our industrialized society. Education is, therefore, the basis of survival for a vastly increased number of people on the planet Earth with its limited resources. Here is a joint task which should not suffer from any quarreling between groups and races.

Co-operation on joint tasks can lead to integration of people coming from different backgrounds. Examples are the integration of millions of refugees in West Germany after World War II, and the continuous integration of immigrants in Israel. Of course, the newcomers in these historical examples were welcome, and they received assistance. In other areas where different ethnic groups live together under tension, with violence expected or already erupted, the members of one group are not welcome to those of the other one. But this is just the point. It is here where new thinking is needed. In the long run, modern technology will lead to close contact between all races, religions and cultures. New forms of living together will have to be developed. We should try to start in those places where different nationalities have lived together for a long time. There we should eliminate tensions as far as possible. But we should not divide people from each other artificially by removing them from their native lands because this leads to uprooting, alienation and to a destruction of cultural values. Under no circumstances should existing social and political structures be dissolved before general agreement has been reached on the new structures which are to replace them, and before these structures work. The prevention of catastrophes must have top priority in technology as well as in politics. No effort should be spared in realistic risk assessment.

In the final analysis, what is needed, is a rediscovery of the ancient ethical command of "love thine enemy". It is an old experience that a perceived "enemy" is disarmed and converted into a friend when he feels treated and trusted as a friend. Conversely, friends can become enemies if they feel mistreated. What is true in the life of individuals is also true, to a large extent, in the political psychology of nations.

What is needed might also be called "tolerance". The ethics of science (not necessarily of scientists) might be of help here. Science wants to discover truth. But science teaches also that any theory about truth developed by scientists on the basis of their discoveries is provisional. It is valid until it is falsified by new discoveries and a better theory (Karl Popper). Science teaches honesty and truthfulness because any lies will sooner or later be detected by other scientists. But at the same time it teaches tolerance towards new, unconventional ideas in the interpretation of the established facts because - who knows? - they might be confirmed by further investigation and become a new basis for understanding some part of the universe or of the human spirit. Every scientist stands on the shoulders of his predecessors to whose ideas he is indebted. To be conscious of this indebtedness means to become modest and cooperative. These virtues are also needed, and justified, in the life of nations whose civilizations, if history is studied carefully, can be shown to be indebted to each other and, in addition, represent the common heritage of humankind.

scientific communities in different nations should cooperate internationally, in working groups, symposia, and conferences, in identifying solutions to the problems which present themselves at the four levels.

To organize this on a regular, continuous basis is an enormous task for the international scientific community and its official national representatives. But if governments are to be induced to stop using old-fashioned methods for dealing with explosive conflicts at the end of the 20th century then our academic institutions should not shrink from the greatness of this task. If they do, the problems that will face humanity will soon be overwhelming. Every possible effort should be devoted to coping with these challenges of the third millennium. Dealing with armed conflicts is one of them. But even without armed conflicts the remaining problems are hard enough. The menace which they represent should unite all of us.

4 A Negative Goal for Science

The prevention of social disasters by interdisciplinary, international scientific co-operation[10]

The intrinsic goal of science is the gaining of knowledge. Members of the human race originally found themselves in an unknown environment. By investigating it, at first probably just for food and shelter, but increasingly also out of mere curiosity, and lately by scientific methods, humans found out that they were living on a small planet of a solar system in one of billions of galaxies in an expanding universe. There is no proof of life in any other place. By scientific methods people also learned to understand, and to exploit, the forces of nature and to use them in the service of human needs and pleasures.

Man, like any other creature on this planet, has always lived at the expense of nature. But in its early history the human species was not numerous enough, and its tools were not powerful enough, for disturbing the overall equilibrium of nature. Only in our days man has acquired the capability, by his technical prowess and due to the great numbers of his species, to destroy the foundations of his own existence, and that of the existence of many other species as well. The other species, in fact, are disappearing first. Man does not have the instinct to preserve nature. When his instincts were formed nature was, for all practical purposes, inexhaustible. It was only necessary to try to avoid direct catastrophes, like being killed by a beast or by a falling tree or by fire, but it did not pay to worry about the secondary effects of any action, as long as that action served the primary purposes of increasing the immediate well-being of the actors.

Today we are still inclined to act according to this habit. We still behave as if nature was inexhaustible. We do worry about accidents and disasters that might be a consequence of failures in our technical installations, like the breaking of large dams, the bursting of nuclear power stations, the collision of ships and airplanes. But even in these cases our precautions are limited by economic considerations. This means that we give priority to short-range

---

[10] Abbreviated version of a paper presented at the International Conference "Science in the Future of Europe", Budapest, 27-28 November 1993. Published in E. Sylvester Vizi (Editor), Science in the Future of Europe, Akadémiai Kiadò, Budapest 1994.

33

advantages, hoping that our present precautions will be sufficient for an indefinite period of time. If the precautions become very expensive we tend to give preference to arguments why they might not be necessary, neglecting arguments which stress the risks involved. An example is the occurrence of shipwrecks of large oil-tankers near coastlines. In spite of a series of bad accidents the international community continues to allow old vessels with poorly trained crews under cheap flags to carry on with their risky journeys although the probability can be calculated when, on the average, the next bad accident will happen by a combination of bad weather, bad luck and human failure. It would be perfectly possible to reduce this probability enormously by requiring, for tankers, double hulls, extra pilots, and stand-by towboats. This would cause additional costs which, however, would be small compared to the environmental damage caused by accidents. Nevertheless, shipowners tend to avoid these costs because governments did not conclude international agreements forcing the shipowners of all countries equally to take effective precautions.

In the examples mentioned the task of avoiding catastrophes is, at least, in the hands of the technical experts for the fields concerned, i.e. for the construction, maintenance, operation and repair of dams, nuclear power stations, ships, airplanes etc. and of the related equipment. If they know their trade, and if they get expert psychological advice on the probability of failures in the reliability of service personnel, they ought to be able to give upper and lower limits for the risk involved in the operation of a given device. It is then a matter of comparing the risk with the expected benefit. Whether the benefit is worth the risk is a decision of values, not a technical decision. Only the educated public and its elected politicians can make that decision.

Unfortunately, however, because nature is no longer inexhaustible and damage done to it may be irreversible, it has become necessary in many cases to take into account also inadvertent by-and after-effects of our actions. They might be called secondary consequences.

Today, most of the large problems facing humanity are to a great extent secondary problems, i.e. by- and after-effects of human activities which had very noble intentions. Environmental problems, in the Third World, are in part a by-product of fighting hunger and, in industrialized countries, of creating affluence and making life more convenient. Overpopulation is a by-product of fighting disease and improving hygiene. Armaments and civil wars are a consequence of trying to improve security and protect sovereignty.

In today's world human activities are entangled in a closely-knit web of interconnections and interactions. Whatever is done in one field affects many other fields. Positive effects in one area may cause negative effects in another area. For this reason operative measures against undesirable consequences of human activities, of primary nature as well as of secondary nature, are hard to implement. The United Nations Conference on Environment and Development, UNCED (Rio de Janeiro, 1992) is a good example. It produced a list of urgent problems regarding the protection of the atmosphere, of freshwater resources, of biological diversity, as well as regarding consumption patterns, demography, human settlement, combatting deforestation, the sound management of biotechnology, of toxic chemicals and of hazardous wastes, to mention only a few problems. Many of these problems are interrelated, all of them would require concerted efforts by governments and scientists on an international and interdisciplinary level, taking into account the economic and political aspects of possible solutions. Solutions are available for each particular problem, but each solution creates new problems in other fields, and some are of particular economic and political significance. Therefore, very often no serious decisions on immediate implementation of the available options are taken, and the deterioration of the environment continues at an appalling pace.

There is a growing conviction that drastic changes will be required in order to cope with the present situation and avoid major catastrophes. These changes may affect lifestyles, standards of living, traditions, political and economic institutions, international relations. As long as all the available options (including the option of doing nothing or continuing business as usual) have not been studied in detail with all their potential consequences (including social, economic, psychological risks and benefits connected with each option) there will be a reluctance to take effective measures. Studies of this nature, however, are difficult because they require the co-operation of many disciplines in science and technology as well as in political science, international law, economy, history, psychology, anthropology etc. Moreover, these studies will have to be carried out in an international framework because the conditions in many different nations will have to be taken into account, and any proposed solution will require international acceptance in order to be feasible. If anybody is suitable for setting up such an interdisciplinary, international framework to study global problems, it is the community of academies of sciences and of national scientific societies.

The tasks to be fulfilled, if rapid progress is to be made toward a solution of the threatening problems, obviously lie on the four levels mentioned in chapter 3:

•   the level of science and scholarship. The knowledge of the origin of problems and of possible solutions still has many gaps which need to be filled by research;
•   the level of practical policy. Options for concrete action have to be worked out, with costs, benefits and risks assessed for each option;
•   the level of the media. Politicians and the public have to be informed about the situation so that the required political measures - even if unpopular at the outset - get the necessary support;
•   the level of education. The young generation has to be kept informed about the state of global affairs so that future leaders are prepared for the tasks they will have to tackle. Knowledge is not inherited, it must be taught and learned.

It is also obvious that the tasks are vast and cannot be fulfilled in a satisfactory, problem-solving manner by isolated efforts, even if such efforts have the size and significance of UNCED. There is a need for continuous, well-organized, consistent, institutionalized efforts if the ultimate destruction of the organic surface of our planet by overpopulation, environmental degradation and wars is to be avoided. International, interdisciplinary task forces should be set up by the community of national academies of sciences and equivalent national societies. These task forces should work, in various fields, on the four levels mentioned above. First of all, they should collect and process the information available, making use, e.g., of the results obtained by the International Institute for Applied Systems Analysis (IIASA), and set up risk profiles, updating them continuously on the basis of developments and of the decisions taken by governments, industry and the public. Estimates, with limits of error, as to how the decisions taken (or the lack of necessary decisions) influence the magnitude of risks should be published regularly. The limits to prognosticability, when too many parameters are involved, should also be clearly stated. At the same time attention should be called to the fact that certain (e.g. catastrophic) events can happen even if, due to the complexity of the situation, it is impossible to predict their occurrence with certainty. This must be taken into account in taking precautionary measures.

Experts on history and on psychology should also be involved. This is important for studying and understanding the obstacles that prevent the implementation of rational measures designed to solve problems. These obstacles are often based on special perceptions of the situation under consideration, and of the resulting priorities. Without understanding these perceptions it is often impossible to arrive at acceptable solutions that can really be carried out.

Obviously, it will be necessary to define and select the "tasks" which the international task forces are to address in an interdisciplinary way. The long-range dangers threatening mankind by the related phenomena of overpopulation, hunger, migration, social instabilities, nationalism and fundamentalism, underdevelopment, environmental degradation and pollution, armaments, civil wars, technical failures, energy shortages, unemployment, overconsumption etc. are well known. Technical solutions are available, in theory, for most of them. But often these solutions are not compatible which each other, and therefore they are not applied. Moreover, bitter conflicts arise because people do not understand each other and misperceive the motivations and intentions of their partners.

Task forces could take stock of the situation in various fields, of the existing trends and the probable consequences if no appropriate action is taken to change the course of events. This could result in different scenarios based on various assumptions.

Options for action should also be drafted which could avoid the most perilous consequences. At the same time, by considering the effects of such action in neighbouring fields, such as national economies and party politics, it should be studied for which reasons the available options for beneficial action are not taken. What are the obstacles? What are the means by which they could possibly be overcome? What is the probable relation of costs to benefits in each case?

To organize the programme described here on a regular, continuous basis is an enormous task for the international scientific community and its official national representatives. But if governments are to be induced to stop using old-fashioned methods for dealing with explosive problems at the end of the twentieth century then our academic institutions should not shrink from the greatness of this task. If they do, the problems that will face humanity will soon be overwhelming. Every possible effort should be devoted to coping with these challenges of the third millennium. The menace which they represent should unite all of us.

(Some parts of this paper have also been presented to the 6th International Amaldi Conference, Rome, September 27-29, 1993; see chapter 6).

5 Is There Any Hope for the Control of Regional Conflicts by Confidence-Building Measures (CBMs)? [11]

The sad facts observable in former Yugoslavia, in the former Soviet Union, and in certain regions of Africa and elsewhere seem to suggest a negative answer to the above question. I think, however, that there is hope. But this hope will only materialize if the individual roots of each conflict are studied carefully so that confidence-building measures (CBMs) can be designed to match the special conditions of the case under consideration. Every armed conflict is based on some sort of enemy images. It must be the declared goal of any CBM to undermine, and finally destroy, a particular enemy image.

Although each conflict is different and, therefore, would require CBMs of a special kind, there are some general rules which apply in almost all cases. Any special CBM must be designed in such a way that it is compatible with these general rules.

Before we specify these rules we must clarify the definition of the "confidence" that is to be built between adversaries involved in a conflict. Confidence with respect to what? Depending on the nature and the degree of development of the conflict the aspired confidence may have different objectives: It may be just the confidence that the representatives of the other side in armistice or peace negotiations are honest, and will keep any agreements reached, even though the enmity underlying the conflict is still in existence. Or it may be the confidence that the other side feels sufficiently deterred by one's own defensive armaments and that, therefore, it will not attack. Finally, and this is the highest form, it may be the confidence that the opposite side in a conflict, in view of the common interests of both sides, is now ready for constructive co-operation. Co-operation may be called constructive if there is the serious intention on both sides to end the conflict

[11]Edited version of a commissioned paper presented at the 44th Pugwash Conference on Science and World Affairs, 30 June to 6 July 1994, Kolymbari, Crete, Greece. Published in Joseph Rotblat (editor), Towards a War-Free World, Annals of Pugwash 1994, World Scientific Publishing Co., Singapore 1995.

in a peaceful way. The confidence that this is so may be based on various assumptions. Either it might be assumed that the other side is now prepared to yield to one's own justified demands, or that the other side is now prepared to accept a compromise that is tolerable for one's own side. It may also be that one's own side has now decided to yield to the justified demands of the other side, and that it is assumed that the other side will believe and accept this.

Whatever type of confidence is to be attained - whenever possible it should be the highest form - the following general rules must be observed if CBMs are to serve their purpose:

- The readiness must exist to interpret the actions of the other side in a positive sense, i.e., these actions must be perceived as being the result of purely defensive thinking. The actions of the adversary in the conflict must be seen as being dictated by fear of one's own potentially aggressive intentions. The Cold War is a historical example: It ended when both superpowers realized that the adversary did not (in any case, no longer) plan to attack, and that the enormous and wasteful armament efforts just served purposes of defence and deterrence (and, possibly, pre-emption), and that both sides would be better off if they cooperated peacefully.
- The readiness must exist to believe that the other side is capable of accepting the priority of common interests over any particularistic interest.

There are cases in which the conditions implied in these rules cannot be fulfilled. Then the building of confidence is not possible. This applies when there is no way to avoid the conclusion that one side is planning the subjugation or even the annihilation of its adversaries, and is using peace negotiations only as a means of deception. Hitler's behaviour before the outbreak of World War II is an example. Under circumstances of this type the only alternatives are capitulation or determined confrontation, with countermeasures including armed resistance.

If the chance of building confidence was either lost by negligence, or did not exist in the first place due to the criminal character of one or both of the adversaries in a conflict, then the conflict will continue. It may end after heavy losses of life and much destruction when a new generation feels that this madness must be ended and when political leaders arise whose strength and prestige allow them to overcome the burden of the past and to create a new partnership for future co-operation. The historical example for this type of a new beginning after centuries of devastating warfare is, of course, the alliance between France and Germany initiated by Charles de Gaulle and Konrad Adenauer.

Since the awareness of common interests is so important for ending conflict and starting co-operation, the remainder of this chapter is to deal with the ways and means to create, in all parts of the world where armed conflicts are raging or are threatening to rage in the near future, a feeling of solidarity and common interest. In past history it was often the fear of a common enemy which induced former adversaries to make peace and co-operate against the common threat. In a beleaguered castle, in the Middle Ages, personal enemies used to make peace for the duration of the siege, and fight the outside enemy of the castle. It was their common interest to survive. When the Germans invaded the Soviet Union, Churchill, an old fighter against communism, immediately offered support to Stalin. He is said to have remarked that if Hitler had invaded Hell he, Churchill, would not have hesitated to co-operate with the Devil. After the Second World War, Western Europe, for centuries torn in strife, united under the perceived common threat of Soviet attack and domination. At the height of the Cold War it was, I believe, President Reagan who said that if the Martians would attack Earth, the United States and the Soviet Union would immediately join forces to fight the Martians and defend Earth. Of course, the forces of the other nations of this planet would join the United States and the USSR in their defensive fight against an attack from outer space.

In a way, this situation has now arisen. The planet Earth is being attacked by a number of global problems, and to defend it must take the uppermost priority over all regional and national problems. The survival of human civilization, and of humankind itself, depends on the joining of forces.

The well-known physicist and elder statesman of science, Victor Weisskopf, named "Seven Cardinal Threats" which he defines as: (1) the pollution of the atmosphere, (2) the pollution of lakes, rivers and oceans, (3) the pollution of the soil, particularly by agriculture, the spreading of deserts and the dying of forests, (4) overproduction and unemployment, (5) problems of the Third World and of the contrast between poverty and affluence, (6) "spiritual pollution" (nationalism, fundamentalism), (7) the population explosion. At the international symposium held in Rome in December 1992 on the occasion of the 50th anniversary of the completion by Enrico Fermi of the first nuclear reactor, Weisskopf challenged the scientific community to devote themselves, by applied research, to these seven cardinal threats.

As mentioned in chapter 4, the United Nations Conference on Environment and Development, UNCED (Rio de Janeiro, 1992) produced a list of urgent problems regarding the protection of the atmosphere, of freshwater resources, of biological diversity, as well as regarding consumption patterns, demography, human settlement, combating deforestation, the sound management of biotechnology, of toxic chemicals and of hazardous wastes, to mention only a few problems. In chapter 4 we discussed the reasons why so often the available options for the solution of these problems are not implemented. Certainly, weapons of mass destruction are not the only threat to international security. Widespread hunger and civil wars in an age of easy communication and transportation lead to instabilities also in areas not originally involved. Ecological degradation, deforestation, destruction of the ozone layer, risks from inadequately designed and maintained nuclear reactors, the rapid progress in the extinction of species, illiteracy in the presence of high technology, drugs and Aids are only some of the additional global problems that contribute to a general feeling of insecurity and dissatisfaction, particularly among young people. Insecurity and dissatisfaction create social instability. Nuclear weapons stored in an insecure, unstable environment are a disquieting thought.

All these dangers of a global nature may be considered common enemies of humankind, like the imaginary Martians, requiring joint action of highest priority. All the traditionally quarrels based on nationalism, tribalism, racism, religion are antiquated and anachronistic when seen in the context of the global situation. A new world order is required, and all men and women of good will have to join in order to bring it about. Good will is necessary but not sufficient. Shrewdness is also required. There are several options for a new world order, and the costs, risks and benefits of each one will have to be worked out carefully, with all conceivable by- and after-effects considered. Scientists have a task here and an obligation.

Samuel Huntington[12] predicted that the 21st century will be characterized by conflicts between civilizations or cultures, just as the 20th century was characterized by conflicts between ideologies and the 19th century by conflicts between nations. No doubt, many of the civil wars that have flared up in recent years may be interpreted as conflicts between representatives of different cultural traditions. In most cases the adherents of these different

---

[12]Foreign Affairs, Vol. 72, no 3, 22-49 (Summer 1993).

traditions had lived peacefully together as long as a strong central authority existed which was recognized by all of them, though perhaps reluctantly, and which kept them together. Considering that in more than a hundred spots on this globe different cultural traditions are maintained by people living in close contact which each other, and that the number and the intensity of these contacts increase due to modern means of transportation and communication, it would be a tragedy for humankind if Huntington's prophecy came true and if a century of cultural warfare were in the offing. Potential theatres of war would be almost everywhere. This must be prevented. The only way to prevent it is either to strengthen and rejuvenate existing central, peace-keeping authorities, or to create new ones, make them efficient and let them gain recognition by all factions of the society for which they have jurisdiction, before the old, outdated authorities have decayed. History offers examples for such peaceful, well-prepared transitions as well as for chaotic revolutions and civil wars with enormous suffering, loss of life and destruction of treasures of cultural heritage.

The latter will happen if the chances for organizing the former are missed. This means that the cohabitation of different cultures on this globe under the conditions of the age of modern communication will have to be organized. Required are institutions with real authority for the prevention or early management of regional conflicts. They must enjoy, or gain by their actions, the respect, trust and confidence of those whose quarrels it is their task to settle. It must become clear that the use of brute force by reckless politicians no longer pays. Disputes must be brought before international courts of justice, and their verdicts must be enforceable, if necessary by an international police force. The use of military force must become a monopoly of the United Nations, to be applied only after due procedure as prescribed by international law, but in a fast and efficient manner.

Again, a historical reminiscence may be in order here. It is not far from the truth to say that during the Cold War the two superpowers had the monopoly of using military force. Within their respective spheres of influence they were able to stifle beginning conflicts and to regulate them according to their interests. Of course, in the Third World conflicts were sometimes stirred up when it could be expected that these conflicts would weaken the influence and power of the other side. When the Cold War ended in the collapse of the Soviet Union the mechanism broke which had controlled conflicts within each of the power blocs. Replacements in the sense of new security structures have not yet been created. Thus, in many places old conflict potentials became virulent which had been suppressed under the old system, and this often led to open warfare. The creation of new mechnisms for the peaceful regulation of regional and national disputes through the United Nations therefore becomes a matter of great urgency. This is the more so since the spreading of weapons of mass destruction is a cause of alarm. Moreover, at the end of the 20th century the repercussions of local conflicts, the flow of refugees, the ecological, the economic and moral consequences are felt worldwide due to the modern means of transportation and communication.

It seems obvious that the world community has to take effective action. A precondition for this to become possible is the creation of a common feeling of solidarity and of purpose. In this respect the institutions of science can make a very useful contribution. They can work out, in interdisciplinary and international collaboration, the social, economic and ecological conditions that have to be expected in the future from the present developments in technology, medicine, population growth, education and social habits. They can offer options for political action, with a quantitative estimate of the costs, risks and benefits for each option. This requires co-operation between the scientific institutions and the governments of all countries, including those countries and regions which at present are fighting each other. This is the type of co-operation for survival which must be enforced in a beleaguered castle. Jointly to be threatened from outside can be an important CBM.

The options for action to be worked out in scientific detail must be compatible with each other. It is well known that they must concern very diverse fields, such as:

- the control of the armaments of groupings at enmity with each other and the design of CBMs fit for overcoming the enmity
- joint measures for agricultural and industrial development with an optimal protection of the environment
- joint measures for an improvement of medical services, birth control and nutrition
- joint measures for education and professional training, particularly of women, with special attention to the need for the maintenance and repair of urgently needed medical and technical equipment, devices and machinery.

But the options must find attention and be acted upon. This will only happen in an appropriate way if joint discussions of politicians, economists, educators, philosophers, psychologists, theologians are institutionalized.

These discussions should not only concern technical solutions for immediate problems. Human beings must have a goal beyond mere survival. They adhere to various philosophies and religions, and have their own ideas on the purpose of life. A pluralistic society must be tolerant. To mediate between the different philosophies and to build confidence among their representatives should be one of the noblest tasks of the interdisciplinary and international discussion circles of this type.

We cannot go into more detail here. Suffice it to repeat that the institutionalization of a permanent, international, interdisciplinary problem-solving capacity is a task of the greatest urgency which the leading national academies of sciences and national scientific societies should accept. If this task were successfully implemented in such a way that decision-makers and the public were ready to take into account the options worked out scientifically, even if this meant making sacrifices now for long-term benefits, this would be the best CBM conceivable, and a big step towards the control of regional conflicts.

6 The Need for Neutral Scientific Advice in Complex Situations of High Risk[13]

6.1   Long-range problems cannot be solved by politicians alone

The president of the Federal Cartel Office in Germany, Wolfgang Kartte, recently expressed the opinion that politicians are overtaxed by events. He accused the German Federal Government and the European Community of being too much engaged with the short-range politics of the day, meanwhile neglecting problems of existential and global importance which require new solutions (Süddeutsche Zeitung, 14 May 1992). Many of these problems are of great urgency. Why aren't they tackled?

In 1986 the Nobel Prize for Economic Sciences was granted to the American scholar James McGill Buchanan for his synthesis of the theories of political and of economic decisions. Buchanan had discovered, inter alia, in his research that politicians, when making decisions on economic policy, were frequently not really guided by macro-economic or socio-economic goals with respect to employment, inflation or economic growth, but rather by the simple goal of winning as many votes as possible and securing their power (Süddeutsche Zeitung, 17 October 1986). Obviously, it is necessary to educate the voter well enough so that the elections will be won by those politicians who present a thoroughly elaborated long-range policy with the goal of a well-balanced, just and stable society, taking into account the by-effects and after-effects of the measures chosen. This means that counter-measures against potential negative consequences are planned in advance.

---

[13]Edited version of a paper presented at the 6th Amaldi Conference "A Contribution to Peace and International Security", Rome, 27-29 September 1993. Published in: Accademia Nazionale dei Lincei, 6th International Amaldi Conference of Academies of Sciences and National Scientific Societies. Report and Documentation, Rome 1994.

6.2   Scientific advice could help. Some examples

This, of course, is more easily said than done, as is demonstrated by the fact that in everyday life well-known risks are ignored over and over again because people prefer to think that accidents won't happen. Even if major catastrophes are threatening in case of an accident, business as usual goes on. An example is the occurrence of shipwrecks of large oil-tankers near coastlines.[14]

The case of tankers is only one example of possible risk reduction by proper preparations based on scientific-technological advice. Further examples can be given not only from technical areas but also from the field of international, national and economic politics.

In chapters 4 and 5 we have already mentioned The United Nations Conference on Environment and Development, UNCED (Rio de Janeiro, 1992) which produced a long list of urgent problems regarding the protection of the environment, as well as a description of the measures that would have to be taken in order to cope with these problems effectively. We have discussed the reasons why decision-makers are reluctant to take these measures, in spite of all their lip-service.

Outstanding among the problems of the world of today are civil wars with their toll of human suffering, material destruction and immense damage to the preservation of ethical and cultural values for future generations. These conflicts have their origins in psychological factors rooted in history, and cannot be prevented, let alone stopped after their eruption, without a deep knowledge of these historical roots and of the psychology of conflict resolution. Here are tasks for leading psychologists and historians specializing in these fields. They could tell politicians that the recipe of trying to solve conflicts by protracted negotiations only makes sense under the very special condition that both sides are interested in negotiated results. A good example was the interest of the two superpowers during the Cold War in avoiding hot war and limiting armaments. Under those conditions negotiations were the only available method for settling disputes, apart from propaganda campaigns, diversion, economic warfare and conventional wars by proxy in the Third World. These negotiations, particularly on questions of arms control and confidence-building measures, often lasted for many years

---

[14]See chapter 4.

and led to meagre results. Nevertheless, open war between the two blocs was avoided. Neither side had wanted war, considering the risk of self-annihilation. Under these circumstances negotiations made sense.

Negotiations are futile, however, if at least one side is not interested in an agreement. Negotiations then may serve only to win time until this side feels strong enough to reach its goals by the use of force, ignoring any assurances given earlier. Hitler's negotiating practice in the 1930s is an illustration of this type of strategy. Another situation in which negotiations will lead nowhere is one in which one side believes it will be doomed if it agrees to stop fighting. Examples for the consequences of a lack of scholarly advice in policy-making may be drawn from the present conflict in Bosnia-Hercegovina. Some politicians try to establish there "ethnically clean" states, provinces or regions for the different ethnic groups. They could have been told by historians and social psychologists that this procedure will not be a road leading to peace. Independent groups or nations which are not united by common interest or common threat perceptions will develop conflicts and even wars unless certain conditions are fulfilled:[15]

- The use of arms must remain the monopoly of an authority at a level higher than that at which the potential enemies are located. Otherwise the war, when started, will probably be fought to the bitter end.
- Parties of the conflict must be made to agree to settling their differences by bringing their cases to an International Court of Justice, and by recognizing as binding the judgements passed by this Court. Its decisions must be enforceable.
- All possible steps must be taken to reduce existing enemy images which are the source of hatred, misunderstanding and fear.

There is, unfortunately, no indication that these conditions will be fulfilled, in the foreseeable future, in former Yugoslavia. Moreover, political scientists and sociologists could tell us that "ethnically clean" nations are obsolete in the age of modern transportation and communication. The trend is indeed in the opposite direction. In our age of high mobility even the old, great national powers of Western Europe - Great Britain, France and Spain - as well as Italy and Germany (united only in the second half of the 19th century) find it

---

[15]See also chapter 3.

Conclusions

Monopoly of force at a supranational level; accepted peaceful arbitration in international conflicts; removal of enemy images; tolerance - all this is said rather easily. It has been propagated many times before, with the well-known poor results. Obviously, it is very difficult to change the traditional methods of dealing with conflicts. Nevertheless, if catastrophes of unforeseeable dimensions are to be avoided in our technological age, it has to be done. How can the findings sketched in this paper be operationalized? The tasks are on four levels:

•   the level of science and letters: the knowledge of the nature of latent and open conflicts, of their origins, and and of the possibilities for settling them by negotiations, mediation, and by the application of international law still has many gaps which need to be filled by research;
•   the level of practical policy: options for concrete action have to be worked out, with costs, benefits and risks assessed for each option;
•   the level of the media: politicians and the public have to be informed about the situation so that the required political measures - even if unpopular at the outset - get the necessary support;
•   the level of education: the young generation has to be kept informed about the state of global affairs so that future leaders are prepared for the tasks they will have to tackle. Knowledge is not inherited, it must be taught and learned.

Scientists and scholars are able to, and therefore have a special responsibility to work at all of these four levels. Of course, some of them, in their capacity as private citizens, might also feel motivated to become active politically. That is what politically gifted citizens ought to do. But occasional work by some motivated scientists is not enough although it is also very important. The work on these tasks must be institutionalized. It must match the high standards of science so that it cannot be ignored. It must supply answers to questions that are answerable.

National Academies of Sciences and equivalent national scientific societies should set up interdisciplinary committees consisting of their experts on psychology, international relations, international law, history, arms control, political science, ecology, nutrition, geography, economy etc. These interdisciplinary committees formed by the official representatives of the difficult today to preserve their national and cultural identities under the impact of massive foreign immigration.

As already mentioned in chapter 3, people are witnessing on TV, in their living rooms, acts of aggression, violence and cruelty in Bosnia-Hercegovina and elsewhere. At the same time they see the inability of the world community to stop these crimes and to bring the perpetrators to justice. This open violation of all rules of decency, with no sanctions following, has dire consequences for the morale of the general public all over the world, particularly for young people (Stephan Wehowsky, Süddeutsche Zeitung, 14 January 1993). Already there is an increasing number of reports on acts of violence in our streets. Victims of attacks do not get any assistance from bystanders who are so used to unpunished violence from watching the news on TV that they lose their moral standards in real life as well. They get used to the role as passive spectator to any act of gross violation of a traditional code of conduct which, step by step, becomes irrelevant. This is very dangerous since a society without moral standards will produce dictators, or will collapse.

Modern society is based to a large extent on a safe application of science and technology. Human failure can lead to catastrophes of Chernobyl dimensions, or worse. Our civilization depends on the availability of well-trained experts for the maintenance of power stations, chemical plants, railway and airline services, oil pipelines and tankers and other facilities. Without them, environmental catastrophes, traffic disasters, electricity blackouts and other malfunctions will lead to a collapse of our industrialized society. Education is, therefore, the basis of survival for a vastly increased number of people on the planet Earth with its limited resources. In spite of this, the quality of our educational system is endangered. There is a tendency, in periods of financial straits, to economize on education and reduce further the already insufficient number of tutorial staff in our universities. To explain the importance of a well-balanced educational system for the stability of our society, and to maintain a "quality watch", is one of the most important tasks of the community of scientists and scholars.

The unification of Germany offers further examples: it was accompanied, or immediately followed, by economic measures like the exchange of East German Marks into West German deutschmarks at a rate of 1:1, and the establishment of the principle of "restitution before compensation" for former private property in East Germany. The consequences of these political decisions for the future economy of Germany were foreseen and predicted by economic experts, as expounded by former Chancellor Helmut Schmidt in his recent book "Handeln für Deutschland" ("Acting for Germany"), Berlin 1993. The forecasts of the experts were ignored by the German Government, with the well-known results.

6.3 Why scientific advice is often ignored

The reluctance of governments to listen seriously to experts is widespread, and not limited to Germany. Most politicians prefer to follow their own plans and to listen to scientific advice only if it agrees with these plans of their own. Advice is unwelcome if accepting it means abandoning a course originally chosen. Why is this so? It is partly the fault of the governments and partly the fault of the experts. The political advice offered by non-political experts is often ignored by politicians for the following reasons:

1. Accepting the advice might require unpopular measures which, in turn, might jeopardize the outcome of the next election.
2. The advice given might have been espoused and advocated by the opposition party. Accepting it might look like yielding to the opposition.
3. The explanation offered for the advice may seem unintelligible, or there may be no plausible explanation accompanying the advice. In this case the suspicion may be justified that the advice given is based on doubtful assumptions or is aimed, in a one-dimensional manner, at just one single goal, without assessing properly the by- and after-effects that cannot be excluded. But often it is just these by- and after-effects which turn out to be of paramount political importance.

As a general rule scientific advisers to governments should refrain from suggesting what should be done unless they are told precisely which goal is to be reached and which boundary conditions are to be observed in approaching this goal. What should be done is a political, not a scientific decision. As a private citizen a scientist may, and should, also give his political opinion on what should be done but he should clearly differentiate between his scientific statement and his political opinions.

When scientists and scholars forecast what is going to happen they should clearly state under which assumptions they have come to that particular conclusion, and what the uncertainties are. In their advice to governments and the public they should limit themself generally to explaining what the options for action are that could be followed, and what the potential risks, costs and benefits are that each option entails. These rules are often ignored so that decision-makers do not feel motivated to consider in earnest the advice given.

On the other hand, scientists and scholars should not wait until they are called upon to give advice. When signs of danger become visible in public developments which seem to go unnoticed by the authorities then it becomes the duty of the institutions of science to raise their voices, point out the facts and their potential consequences and specify the available remedies, if there are any.

6.4   The need for institutionalized international and interdisciplinary scientific task forces

In summary, it might be said that the vast extent and the threatening nature of global and regional problems require intense international co-operation. Close co-operation for joint goals - the goals might be negative ones, like avoiding disasters - can create feelings of solidarity and comradeship and thereby contribute towards the dismantlement of enemy images. This is certainly a desirable by-effect of the joint solution of problems. A prerequisite for the occurrence of this effect is, of course, that the solutions found do not favour, in the perceptions of those involved, one part of the population at the expense of other parts.

Obviously the tasks to be fulfilled lie on the four levels already mentioned in chapters 3 and 4 :

- the level of science and scholarship;
- the level of practical policy;
- the level of the media;
- the level of education.

Specialized international and interdisciplinary task forces for the analysis of special problems should be set up by the international community of national academies of sciences and national scientific societies. A failure of the academic institutions to show concern and to offer advice will, in the long run, lead to the end of academic freedom as we know and cherish it. Scientists will be blamed for the risks which their work has produced for society unless they teach society how to master these risks in such a way that the benefit of their work outweighs by far the unavoidable negative aspects.

7 Arms Proliferation and Nationalism as Parts of the Network of Threats to Peace and Security

The role of fear [16]

Most wars, like most other results of human politics, had a number of different causes. These causes may be classified in different ways, they may be described, for instance, as being of a political, economic, dynastic, strategic nature. Sometimes wars appeared to have started in a fortuitous, accidental way. Whatever the causes, they are always rooted somehow in human psychology. Fear, aggression, lust for power, desire for justice and for military and economic security as well as well as defensive considerations and instincts are involved. These human driving forces are always at work and, of course, they do not necessarily lead to armed conflict. They exert their influence within the framework of a particular historical situation with all its political, social and economic ramifications, and they lead to all kinds of interactions between groups and nations, and often to conflicts of interest. But if such conflicts are not solved by peaceful negotiation and mediation, they may lead to bitter animosities, enmities, and - as a last resort - to war. As Clausewitz has put it: "War is a continuation of politics with the interminglement of other means".

If a particular war is threatening, and if it is to be avoided successfully, the causes of the underlying conflict must be found, analyzed and neutralized in an effective way. To control the supply of weapons to the feuding parties, even to disarm both or one of them completely, will not, in the long run, prevent the outbreak or the resumption of hostilities unless the psychological

---

[16]Paper presented at the 7th Amaldi Conference "How to Reduce Threats to Peace and General Security", Jablonna (Poland), 22-24 September 1994. Published in: Accademia Nazionale dei Lincei, Polish Academy of Sciences, 7th International Amaldi Conference of Academies of Sciences and National Scientific Societies. Report and Documentation, Rome and Warsaw 1995.

reasons for the underlying conflict were removed. The disarmament and isolation of Germany after the First World War is a good example. The Treaty of Versailles had humiliated and disarmed Germany but had not pacified it. Feelings of resentment were a fertile ground for an extreme nationalism. Hitler made demagogic use of these feelings, surreptitiously preparing another war.

Also in our present days, there are feelings of resentment in many parts of the world which contribute towards the rise of a militant nationalism often reckless enough to seek the fulfilment of its goals by military force. Historically speaking, extreme nationalism is the result of war, it spread after the wars following the French Revolution and after the Napoleonic wars. On the other hand, it can lead to war because it is often perceived by other nations as being threatening to themselves.

Psychologically speaking, threatening behaviour of persons can often be explained as the result of fear. This is also true for nations. Self-hate, due to one's own imperfections, is projected onto the others. The fight for justice presents itself with the pathos of hatred. This hatred justifies fear, and it justifies the hatred of the opponent.[17]

Since fear tends to create an aggressive state of mind, in order to compensate for the insecurity connected with it, we must realize that we cannot deal in an isolated fashion with nationalism or armaments (which are just symptoms of fear) if we want to treat threats to peace in a general way. We must look at all the major sources of fear which beset people - politicians, governments and the public - at this time in history. Fear is also produced by distressing situations such as overpopulation, environmental disasters, social inequalities, the gap between the standards of living of the North and the South. There is talk of the necessity that the rich countries will have to use arms to defend themselves against an invasion of hungry masses from the South who might try to find a better life in the industrialized countries. There is fear of unemployment, disorder, social instability and cultural decay. There is fear of Aids. There is fear of xenophobia and "ethnic cleansing", and there is the fear of those who advocate the deportation of foreigners. There is fear of an aggressive Islam. There is fear of clashes between cultures (Huntington). The interconnections between these fears and their links to

---

[17]C.F. von Weizsäcker, Der Garten des Menschlichen. Beiträge zur geschichtlichen Anthropologie, page 475. Carl Hanser Verlag München 1977.

nationalism and to the stability (or, rather, instability) of peace are very solid and cannot be dissolved.

If dangerous threats to peace are to be averted, then the study of methods of arms control though important, is not enough. The whole network of fear-creating problems will have to be tackled in a systematic and coordinated manner.

In the final analysis fear can only be overcome by love, understanding and solidarity. But solidarity cannot become effective without knowledge of the facts that have to be taken into account in practising it. In many threats to peace these facts are very complicated, pertain to different disciplines of knowledge, and are under the jurisdiction of several nations. They must, therefore, be tackled in interdisciplinary and international co-operation.

The present method of handling these matters is, by and large, the traditional way which has led to a multitude of wars in human history. The feeling is widespread that this must be changed at a time when nuclear weapons abound, global pollution reaches a dangerous degree and general misery among three quarters of an ever growing world population is abominable. All possible efforts must be directed towards finding ways of avoiding, or at least mitigating, disasters threatening peace. This needs knowledge, and to a large part scientific knowledge. The scientific institutions whose task it is to collect, screen, teach and preserve knowledge, are certainly called upon to make use of their resources for this purpose. This may need some structural adjustments. The present structures of the organization of science were formed at a time of general optimism with regard to the progress of mankind, based on scientific achievements, at a time when science was not yet widely accused of jeopardizing, by the misuse of its results, the very existence of the human race, at least under civilized circumstances. Today, science must develop new forms of interdisciplinary and international co-operation for finding solutions to global problems. In the past the academies of sciences and national scientific societies were never confronted with a task of this magnitude. It is true that in times of war they devoted themselves to the solution of problems of defence. In "normal" times, however, they usually limit themselves to the promotion of fundamental science.

Unfortunately, the state of affairs seems to indicate that the period in which we live at the end of the 20th century cannot be called normal. The tasks confronting us are of a magnitude comparable to, and even greater than, those which the allied scientific community faced in World War II. A concerted effort is required. International and interdisciplinary working groups of experts carefully selected for each problem or task to be addressed, will have to be set up which permanently follow the developments in the fields of general concern (environmental pollution, population explosion, nuclear proliferation, arms trade, economic aspects of North-South and East-West relations, unemployment in industrial countries, psychological roots of civil wars and xenophobia, deficits in education etc). The institutions of science, of course, have no mandate to tell politicians and the public what should be done, but they are in a position to work out in detail what might happen unless certain measures are taken. They can also produce options for action, specifying for each option the risks, the potential benefits and the costs. Politicians and the public would then have a clearer picture of what the alternatives are from which they can choose. An altruistic option involving sacrifices for the benefit of the suffering should always be among them. As G. Salvini points out, altruism is good business, it is the only good business.[18]

As a first step it will probably be necessary to take stock of the state of knowledge regarding the individual threats to human security in the various fields, and of their interconnections. At the same time a survey could be made of the existing recommendations by expert bodies in the fields of general concern mentioned above.

The next step would have to be to investigate for each of the problematic fields which obstacles are responsible for the fact that many of these recommendations are not implemented, so that the situation continues to deteriorate. It will probably be found that many recommendations did not state clearly enough all the by- and after-effects, for instance in the economic field, that have to be expected if the recommendations are followed. Another frequent weakness is that the recommendations do not discuss the alternative options available for action (including the option of doing nothing), and the by- and after-effects likely to be connected with these alternative options. But only if all the available options are described and compared will it be possible for the decision-makers to choose in an informed way. Only then will the public be able to compare the costs and the potential risks and benefits connected with each of the available courses for action.

---

[18]G. Salvini, 6th International Amaldi Conference of Academies of Sciences and National Scientific Societies, Rome, 27-29 September 1993, Report and Documentation, page 349.

It will be of particular importance to identify clearly the obstacles which so far prevented the implementation of earlier recommendations by scientific groups for dealing with global problems in the fields mentioned above. Was it just lack of public awareness, deficits in spreading information, insufficient effort? Or were some important aspects overlooked (which were intuitively sensed by politicians and the public)? It should be investigated case by case whether the (so far neglected) obstacles justify a modification of earlier recommendations, or, alternatively, what the options are for overcoming the obstacles.

In our age a considerable number of threats to peace is visible, nearby or on the horizon. We must be realistic. Only if the entire network of these interconnected threats is dealt with in a systematic way by interdisciplinary and international co-operation of scientific institutions, offering options for action and pointing out risks, costs and benefits for each option, only then will the discussion of methods of arms control and of the role of nationalism be of real value for the reduction of threats to peace and to general security. The risks involved in nuclear and conventional armaments and the role of nationalism would then be seen in their psychological, political, economic, geographic, historical and other connections with a number of other problems from which they cannot be separated and which become threatening because of these connections.

8 Violent and Peaceful Settlements of Ethnic Conflicts

Some theses on the determinants [19]

Since ancient times the history of mankind has been full of violent ethnic conflicts. The Bible tells us of the murderous wars that occurred in the Near East between 3000 and 2000 years ago. The tales of written or oral history and the legends and mythologies that survived from other parts of the world likewise describe wars and bloodshed between different groups and peoples. Of course, there have been also peaceful settlements and friendly cohabitation. These are mentioned less frequently, probably because they are less conspicuous and do not offer similar material to poets and historians for the description of human heroism and human passion, as tragic conflicts do. There is no doubt, however, that both peace and war are possible within and between human societies. Neither one is inevitable, as numerous historical examples prove. Nevertheless, there is still controversy as to the conditions and actions which are likely to lead to either war or peace when a crisis situation arises, in spite of the fact that quite a bit is known in this respect. Psychologists, historians and behavioural scientists have studied the way individuals, groups and populations behave in situations of stress, excitement, exuberance, depression and despair and have developed theories with respect to the gradual or sudden emergence of enemy images, for instance as a consequence of traumatic experiences which may have occurred in the distant past to members of the group under investigation. It is important never to forget some basic facts of the group behaviour of human beings. They will be described in chapter 12.

Human beings depend on co-operation, therefore they live in groups. They are co-operative towards members of their own group who assist each other in their joint undertakings. They are ready to defend themselves against

---

[19] Invited paper presented at the 17th ISODARCO Summer Course on RACISM, XENOPHOBIA AND ETHNIC CONFLICTS, Certosa di Pontignano (Siena), Italy, 10-20 August, 1995. Published in: Simon Bekker and David Carlton (editors), Racism, Xenophobia and Ethnic Conflict, Indicator Press, Durban 1996.

competing groups. Attack is often considered to be the best defence. Sometimes aggressive behaviour is motivated by fear, sometimes it derives from a feeling of superiority, from the 'right of the stronger group'. This is the law of the jungle. When humans became sedentary their units grew larger. In order to preserve peace internally and organize co-operation and defence it became necessary to formulate and observe laws, to supervise their observance and to enforce sanctions against law-breakers. A leader, king or judge had the monopoly of passing judgement and, on behalf of the community, applying force. The individual was no longer allowed to take the law into his own hands. States developed, and as long as they were stable, internal law and order ruled, more or less. But in dealing with each other the states have not overcome yet the law of the jungle. In modern times this primitive law has been curbed, though quite inadequately, by international law. The bloody wars of the 20th century are proof of this inadequacy. Nevertheless, the United Nations exist. With their 'Blue Helmets' and with the International Court of Justice some small first steps have been taken towards a political organisation which is meant to comprise all of humanity. When it will have been completed foreign policy will be replaced by worldwide domestic policy ("Weltinnenpolitik" according to C.F. von Weizsäcker). The use of force will then be the monopoly of the Security Council or some equivalent body which will see to it that all conflicts are settled peacefully, if necessary by the rulings of an International Court, and the execution of these rulings will be enforced in agreement with world law.

We all know that we are still far removed from this state of affairs. But it is important to know that this is the long-range goal which should be approached step by step. Meanwhile, with the dissolution of the Soviet Union and the disappearance of the Cold War old historical animosities which had almost been forgotten and considered resolved long ago, came to the surface again. Of course, old friendships were also renewed: between the peoples of Eastern Europe and Western Europe, for instance, and between East Germans and West Germans. But also the renewal of old friendships sometimes led to disappointments, because it did not always fulfill the material expectations that became connected with it.

With all the knowledge available on the psychological, historical, economic, and political origins of conflicts it is regrettable that there is little evidence that this knowledge is used by political decision-makers for the avoidance of armed conflicts. Indeed, we do not have to study ancient history in order to find examples of horrible wars and of massacres among the civilian populations, including old people, women and children, belonging to the parties fighting each other. These outrages happen in our present days, can be watched on TV, and are more cruel and bloody then ever due to the existence of modern weapons and technologies. But we should be just: there are also examples of the peaceful settlement of difficult political problems in our days. We shall come to that later.

Obviously, we cannot put up with the present state of affairs when ruthless leaders are allowed to pursue their political objectives by brute force, taking civilian populations as hostages and subjecting them in a most cruel and inhumane way to months and years of continuous shelling and to deprivation of food and medical supplies. Moreover, while doing this these leaders do not have to fear any sanctions. In a conflict of this kind there should have been intervention by superior police forces of the United Nations. Theoretically, the best solution for conflicts threatening to get out of control and turn violent would be, of course, to remove the enemy images on which a beginning feud is based. Can this be done? These enemy images might have their origin in a variety of genuine or merely perceived conflicts of interest, or in racial prejudices, in traditional antagonisms between neighbouring, competing tribes or groups, in imagined irreconcilable religious differences, etc. What is known about enemy images and the methods that could prevent their appearance and growth? Is it possible to "cure" enemy images, and if so, what are the prerequisites for the efficacy of these methods? What can be said regarding the available methods of crisis management and conflict resolution when violent eruptions are to be avoided between groups confronting each other, and when peaceful settlements are to be reached? Under which conditions do social and political changes of the type observable in former Yugoslavia and in the former Soviet Union lead to aggression against minorities, to racism, ethnic cleansing, extreme nationalism, civil war? What role does the weakening, or the fading away, of central authority have in this respect? How do the enemy images originate which underly armed conflicts between populations which, more or less, have lived together peacefully as long as a strong central authority existed? Which psychological tools are available for replacing by cooperative behaviour those enemy images and the racism, chauvinism, xenophobia connected with them?

It would be highly worthwhile to take stock of the available answers to these questions. Of course, this cannot be done in the necessary comprehensive way in this paper and by the present writer who can only offer some ideas and suggestions based on practical experience gained in four decades of work in international relations, both within the academic community and in official assignments in scientific, technological and political co-operation between "East and West" during the Cold War as well as between "North and South". During this period of time many conflicts of varying intensity began, continued and ended, and an attentive observer had the opportunity to notice the accompanying conditions and circumstances. The study of history supplied further examples. The author has learned much by discussions with psychologists, and by reading publications on the results of conflict research.

Dr. Vamik Volkan and his colleagues at the Center for the Study of Mind and Human Interaction (CSMHI) of the University of Virginia have described convincingly that the emotional core of the conflict between groups can often be found in "chosen traumas" and "chosen glories" of the opposing groups:

Chosen traumas refer to the shared mental representations of humiliating events where losses occurred and could not be effectively mourned. Chosen glories recount the shared mental representations of events of success or triumph. Both are often mythologized and passed from generation to generation, although historical hurts are stronger markers of a group's identity than the mental representations of past glories. Losses associated with chosen traumas cannot be mourned and the humiliation and hurt cannot be resolved. Therefore such traumas are handed down in the 'hope' that they can be mourned, resolved, or avenged in the future. This handing down, however, functions to perpetuate feelings of victimization, entitlement, and the desire for revenge rather than successful mourning and resolution. Traumas experienced many centuries in the past are still active in the identities of some groups. It is as if time collapses and feelings about ancient events are condensed and intertwined with current events.[20]

Dr. Volkan and his colleagues at the CSMHI have developed a kind of group intervention process in which leading representatives of the two opposing sides in a conflict meet with a neutral group of experts in psychoanalysis and in the history of the area in question. While the representatives of the two opposing sides attack each other verbally, voicing

---

[20] Methodology for Reduction of Ethnic Tension, and Promotion of Democratization and Institution Building, Center for the Study of Mind and Human Interaction, University of Virginia, 1995.

the accusations underlying the conflict, the neutral observers listen. Later on, they identify and communicate to the two sides the traumas and glories which, obviously, make them think the way they are thinking. A discussion follows. The two sides learn to view the situation not only through their own eyes, but also through neutral eyes, and through the eyes of the opposing side. They learn to understand the 'hidden transcripts' of the power situation between them.[21] Tolerance is promoted, mourning is allowed. Symbolic acts are of great emotional importance. (Examples: Willy Brandt's going down on his knees before the Warsaw ghetto memorial; Mitterrand's and Kohl's joint visit to the Verdun battlefield and soldiers' cemetery.) So is the building of joint institutions.

Of course, before this can be done, the desire for co-operation must be produced. After all, the Serbs and the Muslims in Bosnia-Hercegovina had joint institutions before their country disintegrated. As long as a strong central authority existed they had lived together peacefully in one country. When that country disintegrated they started fighting each other because there was no enforceable legal way any longer which would have allowed them to settle their differences peacefully. The disintegration of Czechoslovakia was peaceful because it was based on negotiations and on treaties recognized by both sides and all neighbours. At all times both sides had governments recognizing each other. Similarly, Russia, the Ukraine, Belorus, Kazakhstan and the Baltic States separated essentially peacefully. The re-unification of Germany was also the result of agreements between legally elected governments of East and West Germany with the consent of Germany's neighbours and of the great powers. There was no bloodshed although the two Germanies had been most loyal members of the two blocs confronting each other during the Cold War.

It seems that one can draw the following conclusions from these examples and from a study of history and of human psychology in general.

Groups of people, neighbouring villages, ethnic tribes, city states, nations or blocs of nations will develop different interests because people are different and have different ideas, philosophies and religious faiths, and are

---

[21] Max Harris, Reading the Mask: Hidden Transcripts and Human Interaction, Center for the Study of Mind and Human Interaction, University of Virginia, 1995.

competing with each other economically, ideologically and politically. If there is a strong central authority monopolising the use of force, and a functioning legal and administrative system, these conflicts of interests will be settled peacefully by arbitration and litigation. If a strong central authority, a functioning legal system and an effective police are missing, fist fighting, murder, mafia rule, civil war and war will start, depending on the size and the armaments of the competing groups. This will only stop if and when one of the following events occurs:

a. complete victory of one side;
b. complete exhaustion of both sides;
c. appearance of a great danger to both sides, forcing them to make peace and unite forces to meet the new challenge.

In the absence of an overpowering central force, conflicts will lead to continuous fighting unless or until one of these three conditions applies. In the world of today a strong central authority is missing, and many wars are flaring up in various parts of our globe. To stop them we have to look for solution *c* because it is certainly preferable to solutions *a* or *b*. History offers many examples for the effectiveness of c. Let me mention a few of them:

- In the Middle Ages, the inhabitants of a castle ended their feuds and concluded a truce when the castle was beleaguered by an enemy force, and the life and freedom of all was at stake.
- Bavaria and Prussia fought each other in the German War of 1866, they became allies against France in 1870, united in the German Reich and never took up arms against each other again.
- For centuries France and Germany waged war against each other. They became friends when both felt threatened by the Soviet Union. Both joined the European Union, and now have joint army units.
- Western Europe, for centuries the site of numerous wars, united during the second half of the 20th century because it perceived a threat to all, emanating from the Soviet Union.
- Eastern Europe was united under the leadership of the Soviet Union and under the auspices of communism against the threat perceived, genuinely or allegedly, from Western capitalism.

- President Reagan, at the height of the Cold War, mentioned that, if the Martians would attack the Earth from outer space, the United States and the Soviet Union would certainly bury their quarrels and would jointly defend human life and values against the invaders.

This means that we have to find some sort of replacement for the non-existing Martians if we want peace among the inhabitants of this planet. Indeed there are great dangers threatening all of humanity:

- The population explosion, hunger, deforestation, the extinction of species, illiteracy in the presence of high technology, drugs and diseases;
- Mental disturbances and instabilities created by fundamentalism and national chauvinism;
- Social consequences of mass unemployment and of the increasing gap between poverty and wealth, both within individual nations and between different nations;
- The vulnerability of our modern scientific-technological society to ignorance and unethical behaviour;
- General pollution of the ozone layer, the atmosphere, of lakes, rivers and oceans, and of the soil; the dying of forests and the spreading of deserts;
- Risks connected with the storage and dismantling of nuclear and chemical weapons, with the illicit production of biological weapons, with nuclear proliferation, with the malfunction of nuclear power stations and with wreckages of supertankers carrying oil.

To master these problems will require the co-operation of the leadership of all peoples. It is an enormous task which needs the good will, the intelligence and the hard work of people all over the world. It will require the construction of efficient mechanisms for the discussion of the available options for mediation and decision-making, with risks and chances assessed, and for the enforcement of rulings by an international court of justice.

Politicians are usually too busy for long-range planning and the comparison of options. They have to deal with the requirements of the day, trying to stay in power by winning the next elections or by eliminating competitors in other ways. Moreover, complicated assessments may be involved which require the knowledge of scientific details and of complex interrelations. This is a matter for the international community of the institutions of science and scholarship. They have access to all sources of reliable knowledge, and their members do not depend on re-elections and therefore can afford more easily to think in terms of long-range solutions, and to publish unpopular results. Therefore, the institutions of science and scholarship should be ready to serve humanity by setting up expert committees for the preparation of options and the checking of proposed solutions for their feasibility, compatibility and their by- and after-effects. Each option offered should be accompanied by a full analysis of the possible consequences, including those to be expected in neighbouring fields. All risks and benefits should be given, if possible, with the probability of their occurrence. The discussion of goals should include ethical aspects, and how to prevent misunderstandings and misperceptions.

The prevention of armed conflicts could be based on a combination of Volkan's three-groups method with a definition of urgent joint goals. Meanwhile, long-range efforts should be devoted to the creation of international institutions of arbitration and peace enforcement which would be universally accepted and respected, except by a few trespassers who would be made to accept them.

The following theses may summarize the *conclusions* of this paper:

- Independent groups of people develop independent ideas which, in time, lead to conflicts of interest.
- If there is no superior authority to which quarrelling parties can appeal and whose judgement is accepted and enforceable, people will feel entitled to use force. They will fight unless there is mutual deterrence of the kind that fighting clearly would mean self-annihilation of both sides. This is true for individuals as well as for ethnic groups or nations. It is the cause of wars.
- Former enemies can become friends and allies when they perceive a common enemy threatening both of them.
- The fading of enemy images can be greatly assisted by studying the chosen traumas and glories (Volkan) of the opposing parties and by making each side understand them. Both sides must learn to see the situation through the eyes of the other side.
- People of different races and different languages can live together

peacefully in one state if they do not feel oppressed and if they respect the institutions of their nation or state. Brazil and Switzerland are examples.

- Peoples which have lived together peacefully under common rule - as the peoples of the Ottoman, Habsburg and Romanov empires - can become enemies when that central rule crumbles (the first two theses then apply).
- If there is to be a dissolution of joint institutions, peace can be preserved if a new, stable power structure is set up before the old one is abolished, and if the transition is made under pre-arranged agreements. Czechoslovakia's dissolution is an example.
- In order to prevent further wars the setting up of strong mediation, peace keeping and peace enforcement authorities should have high priority. Whether these can be developed by the United Nations independently, or with the assistance of NATO, OSCE, European Union or new organizations, is to be studied.
- In the past, joint tasks like hunting a mammoth, defending a castle or building and defending an empire, have motivated people of different origin to join forces and forget their differences. Today, there are global problems threatening the health and life of all of humankind. People should be made aware of the fact that these problems can only be solved by world-wide co-operation and that all the local and national causes for war are minor issues which should be set aside in the interest of joint action for the protection of life on our planet.
- Peace and stability in a multi-ethnic society - and all our societies will become multi-ethnic in the 21st century due to world-wide migration - can only be maintained if education to tolerance and ethical behaviour starts at an early age and continues throughout life, by the school system, the churches, the media, the law system, the trade unions, etc. Without a general consensus on some general and basic rules and principles of behaviour no peaceful and stable society is possible. People must be constantly reminded of these rules and principles. Prominently among them rank the following:
    - honesty
    - readiness to help others who are in distress

- no private use of force; law enforcement only by the central authority
- even a noble purpose does not justify unethical means.

Hans Küng[22] called attention to the fact that there can be no world order without world ethics. In the international conferences of religious leaders organised by him the representatives of the great world religions found that there is a consensus among them on a core of ethics which may be expressed by the Christian commandment 'Love thine neighbour as yourself'. People must come to understand that the highest goal of human life cannot be mere consumption and entertainment (*panem et circenses*). In our fragile world the avoidance, if at all possible, of human and natural catastrophes, and - when they have happened after all - the unselfish assistance to all those who are affected by them must be the highest goal.

Acknowledgement

I am obliged to Dr. Maurice Apprey for reading critically the manuscript of this chapter.

---

[22] Hans Küng, Projekt Weltethos, München 1990.

9 Modern Methods of Information Storage and Exchange in the Service of Problem Clarification, Crisis Management and Conflict Solution

A proposal [23]

Modern communication systems allow access to the largest data banks. Recent developments such as the World Wide Web make these expanding sources of information available to everyone who chooses to get connected.[24] This is very useful for scientists, scholars, medical doctors, journalists and others who are looking for some special pieces of information. To researchers it facilitates the publication of their results which can be made known very swiftly and easily to the world-wide scientific community or to anybody else who is interested.

As a consequence, the idea suggests itself to make use of this new facility for improving the efficiency of tackling the well-known global problems. A selection of them was listed in chapter 8 and, for the reader's convenience, is reproduced here:

- The population explosion, hunger, deforestation, the extinction of species, illiteracy in the presence of high technology, drugs and diseases.

---

[23] Commissioned paper presented to the 45th Pugwash Conference on Science and World Affairs, 23-29 July 1995, Hiroshima, Japan. Published in: Joseph Rotblat, Michiji Konuma (editors), Towards a Nuclear-Weapon-Free World, World Scientific Publishing Co., Singapore 1997.

[24] Caught in the Web. CERN Courier, Volume 35, No. 4, June 1995, page 1.

- Mental disturbances and instabilities created by fundamentalism and national chauvinism.
- Social consequences of mass unemployment and of the increasing gap between poverty and wealth, both within individual nations and between different nations.
- The vulnerability of a modern scientific-technological society to ignorance and unethical behaviour.
- General pollution of the ozone layer, the atmosphere, of lakes, rivers and oceans, and of the soil; the dying of forests and the spreading of deserts.
- Risks connected with the storage and dismantling of nuclear and chemical weapons, with the illicit production of biological weapons, with nuclear proliferation, with the malfunction of nuclear power stations and with wreckages of supertankers carrying oil.

This is just a random selection of major problems that could also be named and grouped differently. Plausible solutions have been suggested for each of these problems separately by expert committees and conferences. However, little progress is made in their solution. On the contrary, in most of the fields the situation is deteriorating. This is probably so because the problems are interconnected whereas the experts usually deal only with one of them at a time, without paying much attention to the interconnections with the other ones. Apart from their direct technical aspects the problems have economic, political, social, historical, psychological dimensions which often are not taken into account in their entirety by the experts. As a consequence, their recommendations meet with obstacles coming from a dimension for which the authors of the recommendations do not feel responsible. The obstacles, however, prevent implementation. Had they been duly assessed in the first place, the options for their removal or circumvention could have been part of the recommendations.

The new communication systems will make it possible to have the suggested solutions examined by representatives of various scientific disciplines, nations, professional, social, political and economic groups. This would allow the decision-makers to become acquainted with all the arguments supporting and contradicting the proposed solutions, and examine them before irreversible decisions are taken. The list of options could be prepared with the risks and chances specified for each option, under short-term aspects as well as under long-term aspects. The frequency of unpleasant surprises following a decision could be greatly reduced if all the potential consequences of that decision would have been assessed before it was taken.

It would be essential, of course, that the input of relevant data to the information system is of the highest quality and reliability. Therefore, it should be controlled by a body universally respected as being committed to unbiassed research and truthful presentation of results obtained. The closest approximation to such an ideal is probably the international community of scientific academies and of national scientific societies. It could develop the means necessary for the continuous supply of the required data to the international information network. This would imply obtaining advice from industry, trade unions, professional associations or whatever section of world society or of specific nations would be considered knowledgeable and competent with regard to the preparation of options for the solution of a particular problem under investigation.

The same principle could be applied when dealing with another widespread and most tragic phenomenon of our time, the outburst of armed ethnic conflicts, civil wars, and wars between newly established independent states. Again, with the assistance of leading historians of the areas in question and of experts of their cultural heritage and of group psychology, data could be stored in the new information system which would show what might happen if certain steps are taken.

Armed conflicts are often based on fear. Fear is often based on mutual enemy images, and the images of the perceived enemies derive their strong colours from the ignorance and the insensitivity with respect to the cultural background and the way of thinking of the assumed opponents. It would be worth the effort to try to design an information system which would allow the exchange of political arguments, producing for each argument chosen the historical, cultural and psychological data supporting it as well as those contradicting it. This procedure could lead to more tolerance and more willingness to accept compromises, particularly if the public is made aware of these data. It could become visible that the stance of "the other side" is not so much based on malice and wickedness but can be understood by considering historical, cultural, economic and psychological facts of life. Compromise may be within reach if proper respect is shown for the facts, both sides treating each other not as enemies but as partners in distress, a distress which must be overcome by joint efforts. Misunderstandings and misperceptions could be clarified and removed by exchanging arguments via the information network. Similarly, objections could be voiced against measures which have short-term objectives without due consideration for highly negative long-term by- and after-effects of, for instance, psychological, economic or ecological nature.

How could the scientific community go about systematically establishing an information system that could serve the purposes described above? The first steps, obviously, should be some sort of stocktaking. There should be a careful analysis of:

1.   the efforts that are already under way to tackle the necessary tasks. (Is this really all that can possibly and reasonably be done?),
2.   the results of these efforts so far and of the available recommendations,
3.   the nature of the obstacles to the implementation of these recommendations:

- lacking power of conviction? Are the recommendations sufficiently well-founded?
- lack of interdisciplinarity? Do the recommendations, for instance, neglect economic, historical, psychological, political aspects or obstacles? Are they based on interdisciplinary research including economists, historians, psychologists, political scientists, etc.?
- gaps in knowledge which could be filled by more research?
- lack of publicity so that the available options and their risks and benefits are not known sufficiently well to the public and to the decision-makers?
- lack of political will because the available recommendations are unpopular due to insufficient explanation of the alternatives?

Put somewhat differently, one might summarize the final programme in the following way:

- What are the great problems of our time which threaten the peaceful development of humankind, thereby creating fear, irrationality, violence and suffering? List them.
- What are the visible solutions? Are the available recommendations compatible with each other? List them.

- What are the obstacles preventing the implementation of a rational strategy for the solution of the problems? List them.
- How can the obstacles be overcome?

Not surprisingly, at each of these four steps the tasks to be fulfilled lie necessarily again on the four levels to which we referred already in chapters 3, 4, and 6:

- the level of science and scholarship. The knowledge of the origin of problems and of possible solutions still has many gaps which need to be filled by research;
- the level of practical policy. Options for concrete action have to be worked out, with costs, benefits and risks assessed for each option;
- the level of the media. Politicians and the public have to be informed of the situation so that the required political measures - even if unpopular at the outset - get the necessary support;
- the level of education. The young generation has to be kept informed about the state of global affairs so that future leaders are prepared for the task which they will have to tackle. Knowledge is not inherited, it must be taught and learned.

Again, for each of these levels we have to study the obstacles and the way to overcome them.

We are entering an age of revolutionary changes in information and telecommunication technology. These changes will deeply affect the lifestyle of people. We cannot ignore these developments which carry great promise. Of course, they can be misused as well. The earlier we get prepared to use them for the common good, the better. We can use the new possibilities of world-wide communication and of general access to the world's stored wisdom and knowledge for the definition and detailed description of humanity's common tasks for the next millennium, for the joint fight against regional and global threats to human beings and to nature, for the preservation of life on our planet.

Conflicts between ethnic groups must be solved by explaining to the parties involved, through disinterested third parties and on the basis of all the historical, cultural and particularly psychological information available,[25]

- the origin of their conflict;
- the reason why each party believes that its case is just;
- the weaker points within these reasons as judged by neutral observers; and
- which possibilities exist for a compromise.

Members of the "disinterested third party" could be appointed by an International Court of Justice. The final judgement of this Court with respect to the type of compromise ending the conflict should be binding and, if necessary, enforced.

It must become public knowledge and conviction that this is the modern way of solving conflicts, based on an intelligent use of communication science and on respect for international law, the only way appropriate for the third millennium. Proponents of the old way of solving conflicts by force must be considered hopelessly old-fashioned and subject to contempt for their primitivity just as a citizen would be who would try to settle his dispute with a neighbour by brute force instead of calling the police or taking a lawyer and going to the courts.

Whereas the new systems for information storage and exchange will be very valuable tools for facilitating, and even making possible for the first time, a rapid availability of all the data necessary for a rational choice between various options for action, and for the settlement of conflicts based on misperceptions and subconscious reactions to undigested grievances of the past, one should be aware of the limitations. The systems can only store, exchange and distribute what has been put in. It has no wisdom of its own. Its successful application will depend on the wisdom of the user. But this remark should not be construed as an excuse for not making an effort to take advantage of the new possibilities for rational decision-making. On the contrary, it is an appeal addressed to all concerned, to develop this potentially

---

[25]See, for instance: Methodology for reduction of ethnic tension, and promotion of democratization and institution building, Centre for the Study of Mind and Human Interaction, University of Virginia, March 1995.

powerful instrument for crisis management and conflict solution, by feeding it with the information needed for the assessment of the available options. This means that it is addressed to the international community of the institutions of science, engineering, political science, psychology, economy, history, indeed of all fields of knowledge that are needed for orientation in the world of today, and for the avoidance of catastrophes which - unless we are very careful and change the present methods of short-sighted decision-making - may very easily lead to the end of human civilization.

10 The Multiplicity of Human Philosophies, and the Threat Perceptions Appearing in the Absence of Common Enemies or Joint Goals[26]

Nuclear weapons are the result of scientific research. There are many discussions nowadays on the role of science in society, and in particular on the role of science in armaments and in the arms race. People do not fully agree. But that is typical for the world as a whole. Disagreement is not surprising. It is the basic problem of humanity. There are different notions as to what the facts are, as to who is responsible for the situation, and as to what should be done. There is no general consensus on any of these questions, except that humanity should be saved. Even that is in doubt in some circles who believe that humanity is not worth saving, and that this planet would be better off without us. Granted, however, that the human race should be saved from becoming extinct, the opinions differ as to how this should be done. The human brain is so complex and so flexible that people can, and do, come up with quite different responses to the same situation.

One example is languages. People isolated from each other for a long time develop languages quite different from each other so that they cannot communicate. Other examples are philosophies, social structures, and religions. Here people who think independently, even if they live in the same environment, arrive at quite different conclusions.

In chapter 2 we used the examples of the ancient Greek philosophers and of the economists of the 19th century for showing that independent thinking can lead to very differing results on the same subject. We found that, in general, the multiplicity of views, interests and traditions results in strife when groups

[26] Edited version of remarks opening the final discussion at the International School on Disarmament and Research on Conflicts (ISODARCO), held jointly by Research Unit Gottstein in the Max Planck Society and the II University of Rome "Tor Vergata", Tutzing (Germany), 20-30 July 1992 (supported by the Volkswagen Foundation). Edited for publication.

representing them get in touch with each other. However, strife and fighting can be avoided if the opponents have come to belong to a social system with a strong central authority that sets rules for a peaceful settlement of conflicts, and establishes enforceable sanctions against violations of the rules. If there is a strong authority or personality or constitution which guides or forces people to adhere to one special way of life, then people will follow that philosophy for a few generations or even centuries, believing that this is the final answer prescribed by God, or by reason, or by history.

Today we are again living in a period of crisis and change of values, of re-evaluation of the past, and of preparing for the future in a new way. Previous crises in history were also connected with horrible bloodsheds, genocides and destructions of cultures. But nature was only marginally affected, and the human catastrophes were always locally confined. So the human race, and nature as well, always recovered sooner or later from the local catastrophes of the past. This time, all parts of the world are interconnected by modern means of transportation and communication, so that repercussions of catastrophes in one part of the globe are felt all over. In addition, the technical means of humankind have increased to an extent that now the self-extinction of the human race as well as of most other species has become technically feasible. We can no longer rely on the healing effect of time, but we have to think in advance on how to avoid catastrophes. Once the catastrophe has occurred, it may be too late for recovery.

What is to be done? We are living in the age of science and technology which have drastically changed the life-style of men and women in the last hundred years by the introduction of electricity, cars, airplanes, radio, television, air conditioning, computers, lasers, spacecraft etc. As in all periods of human history, the military have made full use of all civilian technologies and inventions available in their time. In a few cases, the international community has outlawed certain weapons. The use of chemical and biological weapons was banned, but the arms race in all fields of science and technology essentially goes on. This is not only a great waste of money, but also very dangerous.

Would it help to control science in order to avoid its use for military purposes? The "ambivalence" of science was quoted, and there were some suggestions stating that by public control it ought to be guaranteed that only the "good side" of science should be used. But who would appoint the controllers? In the Soviet Union, science was strictly controlled by the State, but the Soviet Union produced the same weapons that were produced in the

United States where the weapons industry is in private hands. In the West, the controllers would have to be elected democratically as were the politicians who so far decided about the construction of weapons. So, the construction of weapons does not depend on the way science is controlled, or the arms industry is controlled, but on the threat perceptions of governments and populations. What is needed, therefore, is a reduction of tensions by an open discussion of threat perceptions and, secondly, a ban of certain types of weapons including weapons which do not yet exist but which could be designed as a consequence of new discoveries in science. Arms are not the cause but the consequence of hostilities. Men always used all tools available for fighting each other. The only way to avoid a further sophistication of these fighting tools seems to be "preventive arms control", as suggested by Count Baudissin. This means that arms negotiations should start *before* weapons based on new scientific discoveries are built, not thereafter. The "ambivalence of science" is only a special case of the ambivalence of human activities. You can use your hands to feed your children or to kill your neighbour. You can use your tongue to speak the truth or to tell lies and to spread hate and slander. The arms race was not stopped by changes in political systems or by the decade-long diplomatic negotiations. It was stopped by a change in the threat perceptions. The change in the threat perceptions came first. The reduction in arms followed.

In this context we have to repeat what was already said in chapter 2: It must be of paramount importance, therefore, to create a political atmosphere in which governments and populations feel secure, in which no threats are perceived that would seem to justify military countermeasures. In those cases in which old animosities exist, every effort should be made to establish some sort of truce, as formerly between personal foes in a beleaguered castle. In our time, the role of the common enemy necessitating that truce may be assigned to the joint problem of managing survival in the presence of economic and ecological threats of regional and global scopes, cultural clashes, unequal distribution of resources, etc. These challenges, some of which were listed in chapter 2, can only be met adequately by sober, honest and peaceful co-operation. The nations of both Eastern and Western Europe, and as a matter of fact those of all other parts of the world, require a new sense of partnership. Joint institutions for mediation and judgement will be needed. As shown above, truly independent and sovereign institutions will develop different ideas and conflicting interests. There must be a higher level on which such conflicts can be resolved. The decision-making and judgement-passing authorities at the higher levels, that are supposed to stand above the conflicting interests of the lower levels, will require guidance as to which solutions are appropriate. This is a challenge for scientific institutions to form international interdisciplinary committees of scientists, engineers and scholars (including, if necessary, economists, psychologists, historians, political scientists, international lawyers) who devote themselves in an impartial way to the task of developing options for political action in the cases under consideration. For each option, the costs, risks and benefits should be stated, taking into account, if possible, short-term and long-term side-and after-effects. This should include investigations of the potential long-term effects of measures in one field on the conditions in another field. For example, the potential impact of technical projects on the natural and social environments, or the potential consequences of economic decisions for social stability could be studied in depth. Of course, there are questions to which the answer cannot possibly be known, as explained in chapter 2. The future fate of humankind is among the things we cannot know in any detail. There are, however, potential developments that can be foretold because they are the consequence of human actions. The stress here is on the word "potential" because also in these cases it is often impossible to predict what will happen when human actions are involved. But it is often possible to anticipate what *might* happen within a certain range of probabilities. If a major catastrophe is among these possibilities with a non-negligible probability, then it would be irresponsible to assume that this event will not happen. In such cases, the urgent scientific advice should be given to take a less risky course, and the options available for reducing the risk should be described.

11 The Relations Between Industrialized and Developing Countries [27]

Politically, the heritage of colonialism is still felt in the relations between the industrialized and the developing countries. For example, weapons of mass destruction, the possession of which is seen fit for themselves by some of the industrialized countries, are denied to those of the Third World. Another example is the effort by European countries to control and curb the influx of immigrants from the Third World. Economically, it is the debt problem, the problem of tariff barriers and the problem of commodity prices which separate European countries and those of the Third World. Finally, the recent outbursts of violence, particularly in Germany, but also in Britain and France, against people of other races who have set up residence or are seeking asylum in those countries, are indicators of psychological barriers and nationalistic feelings that ought to be overcome.

Europe exists as a community with a special history and a special responsibility. Europeans have to be reminded of that responsibility, particularly with respect to the Third World. Europe is preoccupied now with the problems of European unification and with the difficulties of integrating the countries of the former Eastern bloc, not to mention the problems of dealing with the republics of the former Soviet Union and the war in former Yugoslavia. There is a definite risk that the miseries of the Third World will become forgotten although they will affect Europe's future just as much, or more, as the events in Eastern Europe. Europeans should get together and consider their special role in the global turmoil.

Europe's special responsibility derives from several historical circumstances which are, of course, interrelated: The present situation in many developing countries is deeply influenced by the colonial period during which the European powers determined the fate of their colonies. This is still felt in the fields of politics, economics, education and life-style, is well known and does not need further explication here.

---

[27] Abbreviated version of a paper ("Euro-Sid - is it desirable and/or feasible"), presented to the European Chapter Leaders' Meeting, Prague, Czechoslovakia, September 29, 1992.

There is another aspect of this situation, which is less well understood in Europe. The industrial revolution and the scientific-technological age began in Europe and led to a standard of living there, with which politicians and the general population in the more traditional societies compare their own plight. Television serves to spread information on the European and American ways of life to all parts of the globe. The elites in the Third World are emulating this lifestyle, and ever-increasing parts of the population try to join the elite by obtaining an education of European or American type. In Europe, the United States and Japan, however, where this wasteful lifestyle is widespread, there is already now a growing movement for the protection of the environment against the harmful effects of the consumerism, energy waste and pollution connected with this lifestyle. It is obvious that the global ecology will not, without catastrophes, bear a situation in which a large proportion of the world's population adopted the way of life now considered normal in the U.S. and Western Europe, with two or three cars per family etc. Philosophers like Carl Friedrich von Weizsäcker have called for a change in consciousness, for a new ascetic outlook on life, for a voluntary austerity, for a solidarity of the rich and the poor. Clearly, the rich have to make a start here. The "poor" countries are busy fighting hunger, diseases and ignorance. For them, problems of global ecology can only have second priority.

This means that Europe just as well as the United States, Japan and the other industrialized countries, must set an example and take the leadership in developing a new civilization compatible with sustainable development, allowing the survival of more than 10 billion people on this planet, cultural pluralism and the preservation of nature and of the cultural heritage of mankind. There will be severe conflicts between North and South unless the North succeeds in setting this example. Europe opened the door to the scientific-technological era, which by now has led to a global problem. But it has also led to democracy, to the abolishment of the ancient institutions of slavery and serfdom and to movements for the observation of human rights and for the protection of the natural environment. The affluence of the North represents a special obligation to deal with those problems which, on the one hand, are to a large extent a consequence of this affluence.

On the other hand, this affluence offers chances for tackling these problems, chances which are not available to most of the countries of the Third World. Perhaps it will be the last time in human history that Europe is given the chance to act as a pioneer in the preparation of a new world order, in which the peaceful settlement of conflicts, the preservation of the environment, the fight against hunger, poverty, nationalism and ignorance will be institutionalized.

There is much Europeans can learn from traditional societies. The old values of family and community life, that have become extinct in the hectic consumerism of industrialized Europe, are still alive in many parts of the Third World. They constitute a common heritage of mankind. Just as the heritage of Greek philosophy returned to Europe via the Arab world it is entirely thinkable that the values of individual human allegiance and co-operation, which were developed in many thousands of years of early human history and which were neglected and forgotten in large sectors of industrialized societies, will have to be re-introduced from traditional societies in the Third World where they are still alive. On the other hand, an honest and open discussion of the problems of modern European society, in contrast to those of African, Asian and Latin-American societies, could contribute towards mutual understanding and could assist the efforts towards the design of an equitable global system with a constitution enforcing the peaceful settlement of conflicts.

Thus, no doubt, the Europeans have special tasks and obligations in international development. The results of UNCED could be a suitable basis for further deliberations as to what should be the next steps in the fields of environment and development, and what Europe's role ought to be in taking these steps.

12 Changing of Long-Range Political Goals as a Psychological Pre-requisite for Progress in Arms Control[28]

Introduction

After the end of World War II many observers expected the outbreak of another world war within a relatively short period of time, considering the fundamental differences between the two systems represented by the Soviet Union and the Western nations. The preservation of peace in Europe for more than four decades, so far, is ascribed by many to the deterrent effect of nuclear weapons: The political leaders of both sides realized that war would mean unprecedented destruction.

Deterrence is based on a dilemma, however: How can the probability for the outbreak of nuclear war and the ensuing holocaust be kept extremely low without giving up deterrence? Would it not be sufficient for deterrence, and advisable for reducing the probability of disaster, to keep the number of weapons of mass destruction very small while distributing them carefully to protect them from pre-emptive strikes?

Four decades of negotiations on arms control have led to results which are disappointing when compared with this goal. The present article tries to analyze the reasons for this failure on the basis of the psychological role of group behaviour. It suggests that, if progress in arms reduction is to be made, the opposing sides will have to re-orient their political priorities so that the creation of mutual trust is no longer excluded.

---

[28] Slightly expanded version of a paper PSYCHOLOGICAL OBSTACLES TO ARMS REDUCTION AND POSSIBILITIES FOR CREATING MUTUAL TRUST, presented to the "International Conference of Scientists and Scholars: Can We Prevent World War Three?" (Vienna, Austria, July 26-28, 1985) and published in International Journal of Group Tensions, 1986, Volume 16, nos. 1-4, pp. 105-117. (Expansion carried out in 1985)

The role of group behaviour

Psychoanalysts and students of human behaviour have shown that the instincts of aggression and destruction, just as well as the instincts of love and co-operation, are deeply seated in human nature. They are the result of a struggle for survival in a hostile environment which lasted thousands of generations. In its course man learned how advantageous it was - in hunting, in agriculture, in war - to cooperate with other members of the tribe. A division of labour, a coordination of efforts led to much better results. Now jobs could be done, tasks could be tackled which were completely out of reach for an individual or a small group. The larger a cooperating group was, the greater was its power, given equal levels of cultural development. But in order to keep peace within the group, taboos had to be set up and aggressive instincts suppressed. As *Sigmund Freud* tells us, this suppression leads to recurring eruptions of aggression against those not belonging to the group, such as minorities and neighbouring groups. Therefore, it is not surprising that in the history of human development, from the beginning to our days, wars and violence were permanent phenomena. As groups grew larger and reached the size of nations and alliances of nations, and as technology became more and more advanced, wars became more and more devastating. Today mankind has acquired the technical capability to annihilate itself just by pressing a few buttons.

It is obvious that we must try to re-orient our instincts before it is too late. The old mechanism is still working, of course. In our days two large groups are staring at each other, NATO and the Warsaw Pact.[29] (For the time being I may be allowed to leave the Third World out of the picture.) Each of the two sides is trying to consolidate its camp, and therefore has to project its aggressive instincts onto the other side. Speakers from the Soviet Union tend to see the real fault for the arms race only on the side of the USA, official representatives of the USA see it only on the side of the USSR. None of them asks himself which mistake his own side might have made, e.g. by taking actions which, to the other side, have the appearance of being in disagreement with the verbal statements and offers made. One is tempted to think that it might be useful to introduce into politics the habit of the

---

[29] This was the situation when this paper was written in 1985. The conclusions drawn in the following paragraphs are still valid for any two hostile, apparently irreconcilable, groups of today, e.g. in Bosnia-Hercegovina.

Debating Society of the Oxford Student Union: There a socialist could be given the task to defend the position of the Conservative Party, and vice versa. This forces the debater to understand and to consider the arguments of his political opponents. It would be interesting to see what the outcome would be if at the Geneva, Vienna, and Stockholm negotiations the U.S. representative had to defend the Soviet position, and the Soviet representative the U.S. position![30]

Unfortunately, the mutual projections of fear are made much easier because of convenient ideologies which are at our disposal: one side expects the whole world to become communist one distant day, and feels entitled to help this historical process along wherever feasible. The other side feels obliged, in the name of individual freedom, to prevent just that, and believes that one distant day freedom will prevail world-wide. So each side feels threatened in its very substance, and deeply mistrusts the opponent. This is the old game which makes disarmament virtually impossible. Its motto is: *Si vis pacem, para bellum.*

In our nuclear age, however, the motto must be: *Pax necesse est. Comprehende belli causas!* History shows that wars between former enemies

---

[30] This example is only given in order to show what would be necessary if mutual understanding were to be achieved. Without such mutual understanding it will not be possible to reach lasting agreements. But, of course, one of the problems with this, as Professor *Harvey Brooks* points out in a letter to the author, "is the inhibitions placed on national spokesmen by the domestic constituencies whom they have to satisfy. In the U.S. this manifests itself in the vulnerability to criticism for being 'soft on communism'. In many negotiations with the USSR, the negotiation among domestic interest groups and bureaucratic interests within the U.S. may be more important and significant than the negotiations between the representatives of the superpowers. Satisfying contending domestic interests may place constraints on the position that can be taken by national representatives, however flexibly they may be inclined personally. It is these domestic interests that are less able to put themselves in the shoes of their opposite numbers on the other side of the superpower conflict." Therefore, it is not enough that the national leaders or the official negotiators are flexible enough to understand the positions, interests and goals of the other side. It is mainly their domestic constituencies which have to be educated on political priorities so that, in the long run, criticisms for being "soft on communism" (or, in the Soviet case, "soft on capitalism") lose their political lethality. (Comment added in 1997: In hindsight, one might say that the end of the Cold War and the beginning of nuclear disarmament became possible because *Gorbachev* came to understand that it was *not* the goal of NATO and the U.S. to attack the Soviet Union in a moment of weakness.)

can be avoided if, and only if, these former enemies develop a feeling of belonging to a single group which has common interests. Again, it was *Sigmund Freud* who has shown that man is governed by two competing instincts: one is directed towards destruction; the other is constructive and, when it prevails, leads to friendly, even altruistic, co-operation. The latter instinct is mostly directed towards the members of what one considers to be one's own group, the former to those who are considered to threaten our group from the outside, i.e. "the enemies". This phenomenon which *Freud* describes is being used and has been used by demagogues throughout history to unite their followers by telling them that they have enemies threatening them from the outside. Today it must be our task to develop a feeling that all of humanity belongs to the group with which we identify. There is no longer an "outside". If there must be some target for our destructive instincts, as *Freud* suggests, it should be our own lack of reason and good sense which endangers the continuation of human civilization. The great problems of our time can only be solved in international co-operation. Our aggressive instincts must be sublimed into an urge to work peacefully on these problems. Obviously, the avoidance of nuclear war must have top priority because else all other targets would become meaningless. We must re-assess our ideologies. The Soviet Union must learn that even the spreading of its power, or of socialism, across the world must have lower priority than the maintenance of peace. And we in the West must also learn that nothing can have higher priority than the avoidance of war, not even the defence of freedom, nor the fight against economic recession, nor the strengthening of the Western alliance. It is the other way round: The strengthening of the Western alliance, the fight against economic recession, and the defence of freedom are important for the maintenance of peace, and must be seen as means to that end. The problem for us in the West is, of course, that the USSR also sees the stabilization, growth and extension of its power as something that will serve peace. Both sides tend to justify the application of force in the so-called interest of peace. That is very dangerous. So we must work for a compromise. Both sides must recognize that there is no substitute for peace, and that all other political goals must have second priority. We must make sure that governments in East and West never forget that.

In former times it was the privilege of religious men, of philosophers, of artists, of scientists to work for the whole of humanity. Politicians had a more limited constituency. It extended to nations at most. Today leading politicians ought to take the welfare of all mankind into consideration. Unless they do that, they risk the destruction not only of themselves, and of their own nation, but of human civilization, perhaps of humanity as such. To serve the welfare of all of humanity is very difficult because we do not yet have functioning political institutions for world politics. The United Nations is a step in the right direction, but certainly it does not work sufficiently well. In the long run it will be necessary to establish a mechanism which will match the capacity of the superpowers, and of the transnational corporations, for world-wide action by a corresponding capacity for enforceable world-wide co-ordination.

Any international co-ordination, in its setting of goals, should take into account the historical fact that different nations are at different levels of development and are proceeding on different routes given by their historical past. *Carl Friedrich von Weizsäcker* has called the modern constitutional state with its checks and balances the greatest political invention of the last centuries. It is a rather recent invention. A few centuries are almost nothing in the history of man. Torture was abolished in Europe not so long ago, and it is still being practised in many parts of the world. We cannot expect countries with no previous experience of an honest and efficient administration, or countries in which a large part of the population cannot read or write, or countries which were traditionally ruled by a feudal elite, or by a nomenklatura, to jump successfully into a different system of society which would meet the Western standards of democracy. To expect this, means to think ahistorically. We must realize that it was a slow process for the Western countries to make, and successfully use, what *von Weizsäcker* called the greatest political invention of the last centuries. It seems that conditions of prosperity are required for this invention to function well. Under normal conditions, *Freud* informs us, cultural development, i.e. the formation of larger units of individuals, means restraint of original instincts and is contrary to individual development and to the fulfillment of personal happiness. (I know that *Herbert Marcuse* thinks otherwise but I am not convinced by his arguments. He lived under the affluent conditions of Southern California.)

In any case it will be a slow process for most countries to reach conditions under which a democratic system in the Western sense will work well. In the meantime we must live and co-operate with each other without denigrating each other. We should allow each country to follow its own course, realizing that our kind of democracy may be a close goal for some, and a distant goal for others. The only requirement about which there cannot be any compromise is the prevention of policies which could jeopardize the future of the human race.

Stability and the Sarajevo effect

We all know that nuclear weapons and other weapons of mass destruction, once they have been invented, cannot be disinvented again. Even after complete disarmament the know-how would remain and the weapons could re-appear at very short notice. In other words, mankind will have to learn how to live with these weapons without using them. So far, one might say, the performance of mankind in this respect, since 1946, is not too bad. The weapons were used twice, unfortunately, in 1945, but 40 years have elapsed since then, and they have not been used again by those who possess them. This is probably so because there is no monopoly any longer, and any use of these weapons might result in retaliation in kind. However, there is no guarantee that this type of deterrence will automatically continue to preserve peace. There were, essentially, 40 years of peace in Europe between 1871 and the Balkan Wars which immediately preceded World War I, and there was mutual deterrence at the beginning of this century by the very strong armies and navies of Germany, France, Russia, Great Britain and some others. There was also approximate equilibrium of armaments, as the fact showed that it took more than four years of fighting afterwards until one side prevailed. Nevertheless, World War I broke out. This shows that equilibrium of armaments is no guarantee for the preservation of peace. None of the governments involved at that time had really wanted a world war. But the situation had just become unstable, and the governments did not know how to control this instability. Thus, a relatively minor incident, the assassination of the Austrian heir to the throne, was sufficient to trigger an avalanche of events leading to all-out war.

The situation was different before World War II when *Hitler* was intentionally heading for war without, of course, saying so. Therefore, it is the example of World War I, not World War II, which tells us that an equilibrium of forces and the absence of aggressive intentions are by no means sufficient as preventives against the outbreak of war. World War II can teach us that, under the ancient rules, the absence of adequate defensive preparations will encourage a ruthless aggressor. That is obvious. But in our nuclear age it may be assumed that none of the governments of the nuclear powers of today is planning an intentional nuclear war because that, with a non-negligible probability, would lead to its own destruction. All the preparations for nuclear war which can be seen in the field manuals and in the exercises of the armed forces of East and West are, it may safely be assumed, just meant to make nuclear deterrence more credible, and thereby to preserve peace among the nuclear powers. This gives both sides a good conscience, and that is even more dangerous.

The use of force is not precluded, of course, when there is no risk involved of nuclear retaliation. There are several examples in recent history when nuclear powers were involved in limited wars of this kind. Again, this is dangerous. A miscalculation could have catastrophic consequences. Governments have to be extremely careful to avoid a situation similar to that preceding World War I. It is the "Sarajevo effect", the unforeseen chain of events, of which we must be afraid. Above all, we must avoid the emergence of a state of mind which declares that there are some values in the world which are so holy or so important that they have to be considered to be of higher priority than the avoidance of nuclear war. That was exactly the state of mind in 1914 with respect to the avoidance of war: "If there must be war, let's have it and get through with it!" Peace was considered important, but national honour and economic interests were given even higher priority. This attitude made World War I possible. We must avoid a similar error of judgement.

Let us now look into some of the recent defence policies under the criterion of stability against the Sarajevo effect. Considering the existing distrust, our question with regard to any plan or measure must always be: Does it make peace among the nuclear powers more stable or less stable? And any answer to this question must be critically scrutinized as to the assumptions upon which it rests. What is the probability that these assumptions are correct?

Today we still depend on nuclear deterrence, and nuclear deterrence depends on the second-strike capability of both sides. Any technical development which could be viewed as leading to the capability of either side to eliminate, in a first strike, the retaliatory second-strike capability of the other side to a large extent, would be most destabilizing in a crisis situation: Both sides would be tempted to strike first, in order to pre-empt the anticipated first strike of the other side. Such developments must be avoided under all circumstances.

Defensive strategies

On the other hand, the ideally stabilizing weapon would be one which would be totally effective in defence, and at the same time totally unusable for aggressive purposes. If such weapons existed, and were universally deployed simultaneously, nobody would have to be afraid any longer in any situation of a surprise attack by the other side. Such attacks just would not be feasible. Offensive weapons would become useless. And extensive armaments in such defensive technologies would not have to lead to an arms race because they would be non-provocative. The question, however, is whether such defensive technologies unusable for aggressive purposes are really possible, apart from such devices as the Maginot Line (which proved ineffective in World War II), or as stationary nuclear mines which are unacceptable for other reasons.

In this context let us have a brief look at the Strategic Defense Initiative (SDI) or "Star War" concept.[31] We need not go into the details which have been treated extensively in the literature. Let us only look at the stability aspect here. If it were possible to set up a 100 per cent effective system which would catch all enemy rockets and destroy them before they reach their targets, and if it were possible to introduce that system on both sides at exactly the same time, and if both sides knew and understood this, then such a system would be stabilizing for peace. It would become senseless to plan an attack, at least with rockets, because it would be futile. Offensive nuclear rockets could be scrapped!

So much for the theory. In actual life, however, things look different. The laws of physics and probability show that it will never be possible to make such a system 100 per cent efficient. A certain percentage of the attacking rockets will always get through. The only thing a potential aggressor has to do if he wants to wreak a certain amount of damage on the territory of his enemy, is to enlarge the number of attacking rockets correspondingly. So this will be an incentive for a renewed arms race. Moreover, since the technical capabilities of the two superpowers are not equal in all fields, there will be continuous suspicion on both sides that the other side may be more advanced in its defensive system, or in its capability to penetrate the adversary's defensive system. In a crisis situation of the Sarajevo type, this could lead to the further suspicion that the other side may feel safe enough technically to try a first, disarming strike, and that therefore it would be wise, or indeed mandatory, to pre-empt that strike by a first strike of one's own, thereby at least reducing the expected losses on one's own side. It is clear that the possibility of such considerations in a crisis situation, and the knowledge on both sides that considerations of this type must be going on in the leading

---

[31] See chapter 13.

circles of the adversary, are bound to produce a very unstable situation whenever a severe political or military crisis will occur.

Under stability aspects, therefore, I do not see any possibility for recommending the SDI concept from a practical, non-fictional point of view. Research in this field were only commendable if it were done jointly by the USA and the USSR in order to explore the military potential of new technologies, like lasers and particle beams, for some sort of preventive arms control, i.e. in order to ban certain armament developments before they are carried out. This is not completely utopian. Arms control of this sort has been done before, e.g. with respect to the demilitarization of the Antarctic, and the agreements not to station nuclear weapons on the sea bed or in space.

The need for policies of a defensive character

It seems that for a long time to come it will not be possible by technical means alone to eliminate the danger of surprise attack, as long as each side feels justified in assuming that a surprise attack is among the options held in store by the general staffs of the other side for some future opportunity. As long as that feeling persists on either side, no technical means will be able to grant a sense of security. And unless there is a *sense* of security, there is no security. A sense of security, however, can only be given by political means, not by technical ones. The Swiss Army, e.g., is very strong and is equipped with heavy tanks and fighter planes which could easily occupy or destroy neighbouring cities in Austria, Italy, France or Germany. But none of these countries are nervous about this technical capability because it is politically obvious that Switzerland has its arms for defensive purposes only and will never attack. It is not the nature of Swiss arms which makes the neighbours of Switzerland feel so safe, but the nature of Swiss politics. It should also be the aim of the politics of the USA and the USSR, of NATO and the Warsaw Pact, to make each other *feel* safe. Their relations, of course, are not comparable to those between Switzerland and its neighbours, so this will be much more difficult. But it must be attempted. Military measures of a technical nature are not sufficient. It is true, they can be detrimental or helpful in creating a more or a less tense atmosphere, but they cannot solve the problem as long as mistrust reigns.

Such considerations can also be applied to the stationing of Pershing II missiles in West Germany. It was meant to make deterrence against a limited

Soviet attack on Europe more credible, and it will probably serve that purpose. With the growing Soviet capacity to destroy targets in the United States it becomes indeed doubtful whether a future president of the United States might be willing to risk the lives of millions of Americans for the security of Western Europe. At least Henry *Kissinger*, in his famous Brussels speech, said so, and some leader in the Soviet Union might believe it one day. To counteract this possibility it was suggested by Helmut *Schmidt* and others to station a number of Pershing IIs in Western Europe. They would be on the battlefield if the USSR ever invaded Western Europe, and the probability of their being fired on targets in the USSR in such a case, and before being captured, would be much less calculable for Soviet leaders than for the intercontinental rockets stationed far away from the battlefield in the United States, or for rockets on uninvolved submarines.

So far it sounds plausible: The stationing of Pershing IIs may indeed reduce somewhat the probability of war starting in Europe. But what about stability, and about the "make feel safe" policy? Here the picture looks different. In a crisis situation after an unforeseen Sarajevo effect, when the probability for an outbreak of war looks non-negligible, the presence of these rockets on the potential battlefield results in less stability, not more! When war might be considered imminent anyway, isn't there a temptation for a first strike before these rockets are destroyed by a pre-emptive strike? Might not at least some military leader think so? Might not some civilian activist think so, and take sabotage actions which add to the temperature of an already hot crisis, actions which make ill-considered moves more likely? And is not the stationing of additional rockets with longer range and higher precision a measure which certainly does not fit very well into a policy of making each other feel safe? No doubt these questions, after an honest assessment, will have to be answered in the affirmative. Moreover, people in Western Germany - where the Pershing IIs are being stationed - are also worried for another reason: It may be true that the probability for the outbreak of war is reduced by the stationing of these rockets, but it is not reduced to zero. War might still come some day. In this case it is not only the probability for the outbreak of war which counts, and which must be reduced. The damage done when deterrence fails has also to be taken into account. In first approximation it is *the product of war probability and magnitude of damage* which has to be reduced. For very large magnitudes of destruction - which, of course, become more and more unacceptable - the product becomes non-linear, i.e. the magnitude of destruction enters with a higher power into the product than the probability of war. Using this product-concept one must suspect that the stationing of Pershing IIs will increase the destruction in West Germany in the case of war by a larger factor than it decreases the probability of war. The product would therefore get bigger by stationing, not smaller as it should. It would therefore be advisable, in present and coming negotiations, to aim for a reduction in the number of these rockets, the positions of which would have to be targets for immediate enemy attack, as soon as deterrence failed. The same is true, of course, for Soviet rockets in the GDR and CSSR.

Similar logics can be applied to the "First Use" option on which the NATO doctrine is based. In first approximation it reduces the probability of war because it increases the likelihood that even a war started conventionally will escalate into a nuclear war, and all sides want to avoid that. On the other hand, the probability for a miscalculation increases as time goes by. New generations of political leaders, in decades to come, may have become so used to the existence of nuclear weapons and to the fact that their purpose is *not* to be used that they may start disregarding their lethal potential. They might fall back into the old aggressive behaviour which political leaders have shown throughout centuries and millennia of human history. Their opponents, on the other hand, might be inclined to wake them up by making the nuclear option appear more realistic. Sooner or later catastrophe is bound to happen. The probability for the outbreak of war will never reach zero, it may actually get larger, as time goes by. So we should not only make this probability as small as possible, we should also try to limit the damage as much as possible for the case that deterrence fails. The number of weapons of mass destruction kept in stock should be as small as possible. Considering their destructive potential, and the relative ease with which their number could be increased again at any time, a few of them would be sufficient for deterrence. But the risk for mankind would be much reduced if the numbers immediately available for push-button actions were kept very very low. This, however, may imply the necessity to give up the "First Use" option which *both sides* possess at the present time. An agreement to give up this option should be linked with an agreement to limit conventional forces, in order to avoid a new arms race in the conventional field, and it should, of course, be accompanied by visible redeployments which make early first use infeasible and lend substance to the pledge of "no first use". Mere declarations are not enough.

Possibilities for creating mutual trust

Which possibilities exist for creating mutual trust? In the last analysis this will depend on the notion which each side holds on the long-range intentions of the other side. If it is assumed that the other side, in the long run, will necessarily try to destroy us, the military preparations will have to be different than if we could rest assured that it was only equality and the capability for self-defence which the other side was aiming at. And even if we would have to conclude that it was the destruction of our free democratic system which the other side saw as its goal, or as the goal of the historical process, to be supported by all feasible means, it would still be important to find out which means were considered adequate and acceptable for reaching this goal. In any case, in order to avoid distortions and misconceptions, such as will be produced by projections of our own aggressive instincts, we should learn to see each other through the eyes of the other side. Soviet leaders should understand how Soviet attitudes toward war and peace are viewed by the West. It is well known that Lenin considered a final war between socialism and capitalism unavoidable. It is also known that later leaders of the Soviet Union were keen to convey the impression that, in the nuclear age, Lenin's view no longer holds good, and that other means of competition between socialism and capitalism were preferable and should indeed be used. Nevertheless, the enormous armament effort and the militarization of society in the USSR, with its military training of youth organizations and factory fighting units, show that Soviet leaders think it fit to be prepared for war. What does this mean? Let us now quote some of the questions which observers in the West ask themselves about Soviet intentions:

Are there still believers in Lenin's war theory around in powerful positions? Is it the idea just to deter capitalist aggression, or does one want to remain prepared for the time when capitalism shows a moment of weakness, so that it could be blackmailed without much risk? Is the excessive military effort just a means to impress Soviet allies and to support the doctrine of "socialist internationalism" which means that the way to socialism is a one-way street? Or is it the idea to foster Soviet nationalism and to divert the attention of the population from the miseries of daily life with its shortages of many essential goods? Or do we just witness the inertia of a powerful military establishment which makes sure that it gets all the equipment which science and technology provide, and all the influence it wants?

All these explanations and probably several additional ones can be submitted and substantiated. The truth is probably a mixture of all of them. But it would be important to know the concentration of the different ingredients in this mixture. What is the probability that one or the other of these possibilities becomes official Soviet policy one day? Do we have to be prepared for military adventures? Do we have to brace ourselves for subversive actions of some sort, at home or abroad? Or would peaceful competition between different social systems be possible, particularly also in the Third World, without resort to military force? These are questions which concern political observers in the West. Soviet politicians should know this.

On the other hand, do we know the questions which genuinely worry Soviet leaders? It would be good to be informed. The Soviet Union is a vast country with many resources. This country has suffered terrible losses in two World Wars, during a Civil War with foreign intervention, and from the drastic measures of the Stalinist period. Nevertheless, under the Soviet regime the USSR has developed from a rather backward, mostly agricultural state to one of the two superpowers on this globe. It is still backward, by Western standards, in many ways - particularly in economic respects and as far as the state of personal freedom of its citizens is concerned, including freedom of information and freedom of travel. But there has been progress. For Soviet leaders it is the international status of the USSR, and their own rather high standard of living which count. The Soviet Union and its social fabric are viable, its leadership certainly harbours expectations of further progress. It would hardly be surprising if the elite were worried about signals which could be interpreted, from their point of view, as showing the determination of the United States and its allies to reduce Soviet power and, if possible, wreck its economy, foster dissent among its population and destabilize its social system. Added to these misgivings will probably be questions about the real intent behind U.S. and NATO armament and defence efforts and plans like MX and SDI. Moreover: Who will be the next President? What are U.S., and NATO, and Chinese policies likely to be during the coming decades? Certainly, Soviet analysts will be able to draft a list of questions about uncertainties of Western developments which corresponds to the Western list of questions quoted above. One important psychological difference, however, seems to be that marxist-leninist ideology lends to Soviet politicians a feeling that, in spite of all imperfections and inadequacies in the everyday performance of their system, history is on their side with a very-long-range plan, whereas Western politicians are dedicated to "muddling through" and do not think much about events beyond the next election.

In spite of these differences we must learn to understand each other if we want to preserve peace. Technical measures of arms control and confidence-building are not enough as long as we consider the other side to be our enemies. We must develop a sense of joint venture on this limited planet. Certainly we have different life-styles and different methods, and we may, and should, compete. But each side should respect the right of the other side to be different, and to survive. The aggressive instincts of both sides must be sublimed into a peaceful competition in the framework of the great adventure which the march of mankind into the scientific-technological age represents. Modern information and communication technologies are opening new frontiers whereas problems like hunger, overpopulation, underdevelopment, overarmament are still besetting mankind. Here are many possibilities for co-operation and peaceful competition. Of course, also a peaceful competition has winners and losers. But the loser, in a peaceful competition, does not have to fear for his survival. He knows that he has lost because he was less well-prepared than his opponent, but he also knows that his basic rights will be respected by the winner. On the next occasion the loser will try to imitate the more successful methods of the winner.

We must try to find rules for the East-West competition which allow a cooperative behaviour of this type and which avoid instabilities that could lead to war. This historical competition between the democratic pluralistic countries of the West and the socialist states of the East will certainly continue but it must be carried out in fields and with methods which do not endanger the survival of our civilization and perhaps of mankind.

This means that the following conditions must be fulfilled:

- Every policy, military as well as economic, of both sides, should show the entirely defensive position each side holds. The West should not be reluctant to go ahead here because it is also going ahead in many other respects and often prides itself to have the more advanced social system.
- Each action to be taken should always be seen through the eyes of the other side. Could it be misconstrued by the other side as preparation for a dangerous offensive action? If that were conceivable, every effort should be made to remove such misconceptions.

- Each side should recognize, and respect, the differences in the historical and social backgrounds which distinguish them from each other.

While we in the West often express the honest conviction that our social system is the most advanced one which the world has seen so far, offering economic opportunities and personal freedom and security for a very large majority of the population in our countries, we must admit that the peoples of other parts of the world are living under different conditions and cannot imitate successfully our recipes. They may learn from our experience, but they must build their own roads into the future. We must also admit that our way of living is not only solving problems but also creating new ones which have not yet been solved. Let me just mention waste of resources and pollution of the environment, and unemployment. There is no reason for despair about these problems because they can certainly be tackled by our science and our industry. This is a task for the future. But solving these problems may mean changing our life-styles. So we should not be too proud about our standard of living which is beyond reach for the greatest part of the growing world population. Therefore, it cannot be held up as a model for them. Also we must not forget that our democratic constitutions and institutions needed the preparation of a historical process which lasted centuries. They cannot be transplanted successfully into different environments which have different histories. We must cooperate with people of good will from all walks of the historical process, may they still be living in systems governed by feudal, or *nomenklatura*, elites, or in systems struggling for modern education and for peaceful evolution into the world of tomorrow. If that world is to be a world worth living in, we must leave room in it for many political and social cultures. All these cultures must be considered to be constituent parts of one world culture.

Such a culture would be a natural obstacle to aggression. It would force us to strengthen our intellect and to dominate our aggressive instincts. We should devote our educational efforts to the creation of such a world culture which would tolerate, esteem and encompass all existing cultures as valuable elements. Only if we succeed in this effort will it be possible to create mutual trust and to remove all obstacles to arms reduction.

Conclusions

Let me sum up. Stability against the "Sarajevo effect", i.e. against unexpected incidents which could trigger a chain of events leading inexorably to war, should be the main goal of East-West politics. In theory, a new type of armaments which would be effective in defence but unusable for aggressive purposes, would have the required stability, provided that both sides acquire these new defensive weapon systems at the same time. In practice, however, a situation of this type cannot be expected. The only way out, in the long run, will be to agree on rules under which the unavoidable competition between the different social systems of the East and the West can be carried out in a non-lethal, even cooperative way. Both sides must try to see their own actions through the eyes of the other side.

In the meantime we must behave as if mutual trust already existed, and set up all kinds of joint projects which need mutual trust for their functioning. Some proposals for international projects in, e.g., peaceful space research have been made. But we should not only think of space research. Also down here on earth there are many projects that could be tackled jointly. This way we might set up what in physics we call a self-amplifying circuit. You start with a small amount of trust and some small projects. After a while, as long as we respect each other, stop accusing each other, and see each other through the eyes of the other side, both the trust and the joint projects will get bigger.

Acknowledgement

I am indebted to Professor Harvey Brooks who read the text of this article and made some very valuable comments.

13 SDI and Stability [32]

The role of assumptions and perceptions

Preface

The Strategic Defense Initiative (SDI), ever since it was proclaimed by President Reagan in 1983, has deeply influenced the public debates on arms control, on East-West relations, on internal relations within the Western alliance, on industrial policy, on research policy.[33] It also had a role in discussions on social and political psychology. At the Reykjavik summit meeting SDI was the stumbling block which brought the meeting to an endbefore it had achieved all its goals. At present (March 1988) SDI is still in the headlines. It causes the U.S. Government to insist on a wide interpretation of the ABM Treaty. This is rejected by the Soviet Union. In turn, this rejection causes some leading U.S. Senators to threaten a delay in ratification of the INF Treaty for the removal and destruction of medium range missiles. SDI is also seen by the Soviet Union as an obstacle to progress in the START negotiations on a substantial reduction in the numbers of long-range missiles.

Overall, there has been a tremendous amount of discussion about SDI, both in public and in professional circles, and in political negotiations. So why did we convene another conference on that topic, and why do we publish its proceedings? Isn't the literature on this subject vast enough? There are excellent surveys by competent specialists on many aspects of SDI, such as

---

[32] In this chapter we give an example how scientists, by their special method of asking relevant questions soberly and unpolemically, can clarify controversial questions and uncover the roots of the controvery under investigation. The example is taken from a three-day-workshop held in 1986 with supporters and opponents of President Reagan's "Strategic Defense Initiative". The Text reproduced here is an edited and abbreviated version of Preface, Introductory Remarks and Final Remarks in

K.Gottstein (Ed.), SDI and Stability. The Role of Assumptions and Perceptions. Nomos Verlagsgesellschaft, Baden-Baden, 1988.

[33] K.Gottstein, The Debate on SDI in the Federal Republic of Germany. In: J.Holdren, J.Rotblat (editors), Strategic Defences and the Future of the Arms Race. MacMillan Press, London, 1987.

the volumes "Weapons in Space"[34] and "Strategic Defences and the Future of the Arms Race".[35]

We felt, however, that in most of these many discussions among politicians, scholars, journalists and concerned citizens three elements were missing:

- *Firstly*, a clear definition of what people were talking about. The term "SDI" was being used as a description of very different concepts - let me only mention the "impenetrable shield" of President Reagan's original version which would render nuclear weapons, or at least nuclear ballistic missiles, obsolete and impotent, and, on the other hand, the almost conventional ideas of terminal defense of silos which would just enhance deterrence. So some clarification was needed. Too many different notions and weapons systems had reappeared under the heading of SDI. Many of them were very old and were under development long before President Reagan's SDI speech.

- *Secondly*, a systematic approach was lacking. In most cases people were just concerned about the feasibility of various SDI concepts, or about the economic aspects of the programme , or about the impact it would have on deterrence or strategic stability, or about domestic aspects, or about the attitude of the European allies of the U.S. towards the Programme , or about the consequences for arms control and East-West relations. But hardly anybody looked at the system as a whole, considered the cross-relations among all these fields, and worried about the overall balance of advantages and disadvantages. For example, some measure which was technically feasible, economically desirable, and strategically stabilizing under the assumptions of a continuing confrontation could still be destabilizing politically for psychological reasons , perhaps just because of the underlying assumption of a continuing confrontation.

- The last remark already points to the *third* missing element of many SDI papers and discussions. It is the problem of perception: Usually we judge from our point of view whether something we are planning would have a stabilizing or destabilizing effect. But that is an insufficient way of analyzing our options. We should also look at them through the eyes of the other side

---

[34] Franklin A. Long, Donald Haffner, Jeffrey Boutwell (editors), *Weapons in Space*. W.W. Norton & Co., New York, 1986.

[35] J.Holdren, J.Rotblat (editors), *Strategic Defences and the Future of the Arms Race*. MacMillan Press, London, 1987.

and determine how our actions must appear to our potential opponents. We are convinced, e. g., that our side will never strike first and that defensive measures of the SDI type can only be seen as an attempt to defend ourselves against a potential first strike of the other side. So the supporters of SDI see SDI as stabilizing assuming that also the Soviets know that they will never be attacked if they remain peaceful, and that SDI is only meant to protect U.S. second-strike capability. Seen through the eyes of a Soviet observer who suspects that a first strike might be among the options for future U.S. policy, the situation must look different. To him SDI could appear as a protection against a retaliatory second strike by the Soviet Union. Therefore, in a crisis situation a defensive system against ballistic missiles might destabilize the situation further because its presence might enhance the fear of a first strike and increase the temptation for pre-emptive measures. It is this sort of psychological reasoning which has been neglected so far to a large extent and to which our Workshop was to call some attention.

It was the main purpose of the Workshop to clarify the differences in the assumptions and the differences in the goals and value systems which apparently had led to the differences of opinion about SDI.

One of the main results of the discussions of our Workshop was that the judgement on the (positive or negative) effect which SDI is likely to have on strategic stability as well as on crisis stability depends on our assumptions or implicit perceptions of the goals of the other side. This result found expression in the title of this chapter. The Workshop on which it is based was convened under the heading "U.S.-German Workshop on the Political, Strategic-Operational, Economic and Psychological Aspects of the Program to Investigate Possibilities for Strategic Defense (SDI)". It was held on the premises of the Evangelische Akademie at Tutzing, near Munich, December 14-18, 1986., under the auspices of the American Academy of Arts and Sciences and of our Research Unit of the Max Planck Society, and was financed by the German Research Society (Deutsche Forschungsgemein-schaft). The idea for the Workshop originated when the author visited the Center for International Studies at the Massachusetts Institute of Technology (MIT)..Some of his colleagues there pointed out that in the United States SDI was regarded almost exclusively from the point of view of American interests. They reported a lack of attention to the European viewpoint, let alone to Soviet perceptions and the consequences thereof. For this reason it would be desirable, they said, to discuss these questions at a conference in Europe between German and American experts. For this reason it was proposed that a U.S.-German meeting on SDI be organized to deal with those questions which have been neglected or put aside in the public debate.

In order to be able to discuss these questions in a realistic way an overview of the technical aspects of SDI had to be included in the programme. It was possible to fulfill this requirement in an excellent way given the fact that our Workshop was attended by, among other experts, the former Chief Scientist and Deputy Head of the SDI Organization in the U.S. Department of Defense (DOD), and by two former directors of the Advanced Research Projects Agency, DOD. Thirty-three scientists and scholars, diplomats, military officers, administrators and representatives of industry attended the meeting, among them eleven Americans.

In order to prepare the agenda we had a three-day preparatory meeting at Ringberg Castle in March 1986 which was attended by twelve scholars including two Americans. A number of basic questions were identified which underlie the controversy over SDI. Among them were the following: Is SDI offensive or defensive? Does SDI increase deterrence or do away with it? Is SDI to be kept secret or should it be shared? Does SDI assume a clear distinction between aggressor and defender? Would non-participation in SDI lead to severe economic disadvantages? Does SDI link Europe and America or divide them? Will it be possible to end the SDI Programme once a large investment has been made - if it turns out during the research phase that the Programme is unwarranted for reasons of stability or security? What kind of spin-off can be expected from SDI? What effect on the economy may be expected from the channeling by the government of large resources into specific directions?

A detailed discussion of these questions was published elsewhere.[36] The answers to these questions depend, of course, on various assumptions, presuppositions and perceptions. We decided, therefore, that it would not be the purpose of the Workshop to come up with definite answers to these questions. If participants would offer answers at variance with each other, we would not try to conclude who is right and who is wrong. We would rather try to identify the different assumptions and premises which lead intelligent and honest people to answer these questions so differently. We would then

---

[36] Klaus Gottstein, An Impartial Look at SDI. A Preliminary Report. In: Strategies for Europe. The Link with Technology and Arms Control. Forschungsstätte der Evangelischen Studiengemeinschaft (FEST), Heidelberg, 1988.

look at these assumptions and see how they fit together and what they mean for neighbouring fields. For example, how do the strategic assumptions relate to the political assumptions, and what is the connection between the assumptions of the public as expressed in domestic politics, and the assumptions of the technical experts? We would ask ourselves what the options are, and which assumptions must be made for assessing the corresponding risks and benefits, taking into account by-effects in neighbouring fields, and long-term effects. If it turns out that acting under this or that specific assumption would mean undergoing major risks, we should ask ourselves: Is there a way to change that assumption? Are there several options how to do this? Would some of these options, by avoiding the risks we have recognized, lead to new risks in other fields? What is the most suitable kind of cost-benefit analysis to compare these risks and to choose the most beneficial, least risky path for action? What criteria are to be used here?

This kind of impartial, systematic, scientific approach allowed us to group together experts from different backgrounds and with different views on the desirability or undesirability of SDI. We succeeded in having very fruitful discussions following the invited papers. We did clarify the assumptions and premises leading to the different stances on SDI. Estimates were made of the risks and benefits with respect to strategic and crisis stability. Even though the programme was perhaps not completely fulfilled so that there is room for further research, the editor, for one, has the impression that the Workshop succeeded remarkably well.

At the meeting we did not try to summarize the results obtained. There was general agreement, however, that these results are highly significant for the continuing debate and for the political decisions to be taken in the near future. Depending on special background, previous experience and political inclination, different readers will find different pieces of information in this volume of particular interest. Any summary or abstract, by necessity, would have omitted something that might have been especially important to someone with respect to a particular point of view. Some significant results and clarifications came out only during the discussions. How strong or weak an argument is shows up when it has to be defended against expert criticism. Among the findings which the present author considered particularly noteworthy were the following:

- Any SDI system realistically conceivable will be able to handle not more than 200 arriving re-entry-vehicles. Therefore, a limitation of numbers by negotiations would be a necessary requirement for a functioning SDI system. This requirement is contrary to the original SDI concept of President Reagan who wanted to make the security of the American people independent of the good will of the opponent. On the other hand, reducing the number of offensive missiles on both sides to a figure below 200, and protecting these missiles against elimination by a first strike, could lead to a stable situation in which no incentives for a first strike can be perceived.

• Clear-Cut protection against a first strike does not negatively affect stability, but protection against a second strike does. The problem is that the two kinds of protection cannot easily be distinguished from a technical point of view. Crisis stability, therefore, depends on the mutual perceptions of the intentions of the respective opponents : Is it safe not to assume that the opponent is preparing a first strike and seeks to protect himself from retaliation by defensive measures? Could it be that the opponent is afraid of a first strike and might be tempted to pre-empt it? Or can he be trusted that he will never attack first, under no circumstances? So what counts is not only the intentions of each side but also the perception of these intentions by the other side. In other words, the only solution to this psychological dilemma seems to be the creation of trust. Apparently, there cannot be any significant progress in arms control and disarmament as long as mistrust about the long-range intentions of the other side continues. Sakharov said something similar when he called for ideological compromises between the two systems. As long as we consider each other as ideological arch-enemies, there cannot be any progress. We have to understand each other's intentions and accept them in principle as justified and tolerable. This may mean that some of these intentions and some ideological doctrines will have to be altered.

The meeting reminded us that behind the foreground-arguments of strategic balance and East-West arms control there are also a number of less conspicuous ideas which to some people, however, provide the main motivations for supporting SDI:

- insurance against accidents
- protection against irrational leaders like Qaddafi or Khomeini
- creating of jobs
- breaking the Soviets economically so that they give up their military competition with the U.S.

If SDI is so useless, or even detrimental to the interests of the West, as the critics say, why do the Soviets oppose it so vehemently? In condensed form the answer given was: The Soviets oppose it for political, economic, and military reasons. SDI forces them to take countermeasures. They will have to investigate it, imitate it, look for weak points, build weapons to defeat the system. That binds resources. An economic objection from the Soviet point of view might be that the USSR prefers to avoid, for economic reasons, competition in the fields of offensive forces, defensive forces and in the defeat of defensive forces.

Introductory remarks

SDI is meant to make peace more secure. Whether SDI has the potential to do this or not, is not certain as yet. It is a matter for research the outcome of which is still an open question. This ist also conceded by proponents of the SDI. While this research is going on, however, it has already enormous effects. It influences not only the economy in the countries connected to the programme but also East-West relations, the internal relations within the alliance, the strategic thinking of the military, and, quite generally, the psychological situation of the public in East and West, and, thereby, domestic affairs. To investigate these effects is just as important as it is to do research about the technical feasibility.

Even if research should show that the technical feasibility is rather limited the research programme as such will have had its political effects which we have to take into account and for which we should get prepared. And if research demonstrates the feasibility of certain technical solutions we should find out about their political and strategic implications. In particular, we should identify the conditions and circumstances under which these technical solutions, and possibly later the deployments, would have a stabilizing effect, and those which would lead to instabilities.

How realistic are the assumptions about the likelihood of the occurrence of conditions and circumstances under which SDI would have a stabilizing effect, and of those under which instabilities and a general worsening of the international situation would result?

The answers to these questions will be of great importance not only to the United States but also to its allies in Europe. In the Federal Republic of Germany in particular there has been a hot discussion on what SDI means for the strength of the alliance, for East-West relations, for European security, and for the stability of peace. This discussion is still going on. It influences the relations between the U.S. and Germany. When I visited the U.S. last year and this year and took stock of the situation with some colleagues both inside and outside Government, we had the impression that there was some need for an interdisciplinary, bilateral discussion of the relevant issues. Usually these issues were discussed from the point of view of one nation only, particularly in the U.S., and also from the point of view of one discipline only. The military would look at it in a way different from that of the politicians, or the diplomats, or the industrialists. With a few notable exceptions there were no attempts to look at the national and international scenes as a whole and determine the impact of SDI on the entire system. Of course this is very difficult but we felt that it should be tried. Too often in history and politics the intended effects of actions taken were overridden and spoiled by side-effects which had been overlooked because of linear thinking instead of multi-dimensional or global thinking.

How are we going to run this meeting and what is its purpose? As just mentioned in the preface, we shall try to identify the assumptions underlying the different opinions. We shall then look at these assumptions and see how they fit together and what they mean for neighbouring fields. For example, how do the strategic assumptions relate to the political assumptions , and what is the connection between the assumptions of the public as expressed in domestic politics, and the assumptions of the technical experts? We shall ask ourselves what the options are, and which assumptions must be made for assessing the corresponding risks and benefits, taking into account by-effects in neighbouring fields, and long-term effects. In doing this we have to apply psychology. The psychology of the potential opponent must be considered just as that of our population. It is a result of psychoanalysis that illusions which veil or deny helplessness finally lead to hopelessness, and often to suicide. Could it be that SDI is an illusion veiling helplessness? Do we have to prepare for this possibility? In any case, our assumption in running this workshop will be that all participants are intelligent and honest, so that there is no need for polemics. We shall assume that all views expressed are acceptable if judged within their own frame of reference, and if they are different it is because their frames of reference are based on different assumptions. So it is these *a priori* assumptions which have to be examined.

As an illustration for this kind of treatment let me give you an example. At this workshop we shall hear arguments for the opinion that SDI enhances deterrence by complicating the planning for a first strike, thereby making the strategic balance more stable. This opinion is based on the assumption that the Soviet Union knows as well as we do that the West will never strike first, so that SDI is *not* meant as a protection against their *second* strike. It is to be a protection against a first strike. So it further assumes that the Soviet Union might consider striking first and has to be deterred from doing so. Under these assumptions the opinion on the stability-enhancing effect of SDI is justified. However, whether stability is increased or decreased by SDI depends not only on our perception of the situation but also on that of the Soviet Union. Here we have to distinguish, in theory, the following four alternative cases:

		Warsaw Pact	
N		defensive	offensive
A	defensive	c	a
T	offensive	d	b
O			

a. This is the case I just mentioned: The Soviet Union is considering the option of military aggression against the West - at least in the long run - and believes at the same time that the leadership of the West can be relied upon as being totally peaceful and defensive toward the Soviet Union. In other words, the Soviet Union has the option of striking first but the West will never do so unless provoked.

b. The Soviet Union is considering the option of aggressions as in a. but believes that the West has the same option and that the Soviet Union must be prepared for deterring, or even repelling, what the Soviet Union would consider to be an unprovoked attack.

c. The Soviet Union is excluding the option of starting a major war itself and is also convinced that the West can be relied upon as excluding the option of military aggression for future decades.

d. The Soviet Union is excluding the option of starting a major war but is not convinced that it can afford to neglect completely the possibility of being attacked by the West.

How realistic these four cases are is a different question which has to be investigated separately. It would not be prudent, however, to decide off-hand that any of these four possibilities could be left out of consideration. It should be mentioned that cases *c.* and *d.* do not necessarily mean that the Soviet Unionwould have to give up its ideology of a continuing historical struggle between communism and capitalism in which the latter will be defeated in the long run. They would only mean that the option of outright war would be dropped. The struggle might still be continued by other means like so-called liberation movements, subversive groups or legal political organizations, but against means of this kind military or technical measures of the nature of SDI would be ineffective, or even counterproductive.

So the assumption underlying SDI and the stability-increasing effect of SDI is that case a. represents reality. The Soviet Union, on the other hand, seems to assume that case d. is true and that SDI is meant to protect the U.S. from a retaliating second strike of the Soviet Union, thus making a first strike more likely. In fact, in a situation in which both sides distrust each other of being aggressive a defensive shield in the hands of the potentially aggressive opponent is a matter of concern. It might give a feeling of security to the potential aggressor and make him more inclined to consider attack an option in a seemingly favorable situation, or under stress, than he would be without that protection. At least the defender might think so. Therefore the defender might start investigating the idea whether it would not be in the best interest of his own population to pre-empt an impending attack by a strike of his own. The net result of all these considerations would be that a situation of the type *a.* or *d.* moves to be of type *b.*, with both sides harboring offensive options. This would mean less stability in a crisis situation.

What we have presented so far is still an oversimplified picture because we did not take into account the perceptions of the positions of each side by the other side. Take, e. g., case *a.*: We know that NATO is defensive and will never strike first. But it would be irresponsible not to take into account the possibility, if not the fact, that the Warsaw Pact does not take this for granted. Otherwise we could rest assured that the military leadership of the Warsaw Pact, knowing the defensive stance of NATO, will not consider it necessary to contemplate, and prepare, pre-emptive measures. Unfortunately and most probably, this is not the case. We have to be prepared for the possibility that our defensive intentions are not perceived as such. This means that real life is more complicated than the four clear-cut cases of our diagram. In reality there are no sharp boundaries between the four cases. In our example: We may consider NATO as purely defensive; but since the Soviet Union does not necessarily share our perception we must be prepared for a situation in which the Soviet Union takes measures as if NATO had options for aggression. The realities of political life, as they are revealed by the military and political precautions taken by both sides, are not described by the pure case *a.* but by a situation somewhere in between *a.* and *b.* This, of course, is a description by a Western observer believing in the purely defensive character of NATO and also believing that the Warsaw Pact cannot be relied upon as being purely defensive. We have to deter them. By analogous lines of thought Eastern observers believing in the purely defensive nature of the Warsaw Pact - but being realistic about Western perceptions - would place the real situation somewhere in between *d.* and *b.*, because such observers know that NATO suspects the Warsaw Pact to have aggressive options.

The only stable situation would be *c.* in which *both* sides have only defensive intentions, and recognize this. It would be a case like Sweden and Norway, or Switzerland and Austria, or Canada and the United States. A sudden attack by one of these countries against the other one, its neighbour, is considered unthinkable. There is no need for deterrence against an aggression by this particular neighbour. The situation is stable in this respect, by any stretch of the imagination.

As we have seen, the area of stability, *c.*, is not touched by the scenarios describing the relations between NATO and WP. What can be done to get there? As compared to the present situation it would even be progress if case *c.* would be considered a possibility by the planners of both sides, an option to be taken into account. The case that both sides are perceived by both sides to be non-aggressive must become realistic if stability is to be achieved. During this workshop we shall have to ask: What is the role of SDI in this connection? Does it move us toward the region of stability as it is intended to do? Or does it move us away from it? There are supporters and followers for both these opinions, also at this workshop. We shall listen to their arguments, and analyze the assumptions they use. Our goal will not be, let me repeat, to say who is right, and who is wrong. We shall just examine the assumptions - explicit or implicit - which they are using.

Final remarks

The purpose of this workshop was to clarify the differences in terminology, the differences in the assumptions and the differences in the goals and value systems which apparently have led to the differences of opinion about SDI. We have been able to clarify the various notions of SDI. Apart from identifying SDI I and SDI II we have seen that there are also several varieties of SDI II. When some people talk about version II of SDI, they see a full-fledged programme of interception in all phases, boost-phase, post-boost-phase, mid-course and terminal intercept with efficiencies anywhere between 10 % and around 90 %. According to them, that is the goal of the SDI program. Others think that an insurance policy with only mid-course interception and terminal interception is already sufficient, provided you get some sort of arms control agreement that will limit the number of incoming re-entry vehicles. But a defense consisting of terminal intercept alone might be the most cost-effective way of doing the job. It was mentioned that terminal defense is an old programme that wouldn't have needed SDI for its execution. Nevertheless, under SDI all kinds of new technologies are made available for this job, too.

It also became clear that nobody among the experts believes in SDI I, the original vision of President Reagan. But the view was expressed that the public, nevertheless, must be told that SDI I is still the goal, although perhaps a goal for the very distant future. Otherwise the programme wouldn't survive. The problem with telling the public something in which the experts don't believe, is, of course, that this is not honest. There is always a risk with dishonesty, namely that it will be discovered some day. Then the public will lose trust in the experts and politicians, and the possible result of this losing of trust could be a further increase in the already existing enmity in some circles against technology, further doubts about the leadership of the U.S. where such dishonesty would have originated and an undermining of trust in our own government which would have been a partner to this dishonesty. I would say, it's better to be honest and confess that SDI I may be a nice dream but that realistically only SDI II can be expected in one of the varieties just mentioned.

Of course, the public would have to understand that this sort of SDI is not a protection for the population but an enhancement of deterrence. But that's what it is. It could still be a worthwhile project if it led to a reduction of offensive arms. In fact that is what was said would happen if this programme were carried out. Actually, it is necessary for this to happen, even for the SDI II versions, if they are to have a reasonable effectiveness. More than 200 incoming missiles cannot be handled very well. Any greater number must be reduced to about 200, either by shooting them down in the initial phases of their flight, but that is much more difficult than intercepting them in the final phases, or their number must be limited to about 200 by arms control agreements. But what are the assumptions which have to be made if we are to believe that this will happen? Obviously, we have to assume that the Soviets truly believe that we shall never threaten their second-strike capability by a first strike. Only then can we expect them to react to an improved ABM system by a reduction in offensive weapons and not by an increase. So far, at least, they make statements that they will react to SDI with an increase in offensive missiles. So they do not seem to trust the entirely defensive intentions of NATO. or if they do, they do not admit it. The reverse is also true: We do not trust their defensive intentions. Secretary Weinberger said that he would not sleep very well if the Soviets had a well-functioning SDI system and he also said they are working on that.

So, both sides expect the other side to work for an SDI system in order to protect against a retaliatory second strike of the other side. Such a system would, of course, improve the first-strike capability of the other side. Thus, in a crisis situation both sides would become even more nervous, and crisis stability would be decreased under this mutual perception of the situation. However, for SDI to work sufficiently well and to fulfill the President's hope for reduction in offensive arms we have to assume the opposite: We have to assume that both sides trust each other at least to the extent that they do not see the SDI work of the other side as a protection against a retaliatory strike but only as an insurance against a first strike. So SDI would be very good for stability as long as both sides concede to each other that they are not planning a first strike but just want to buy insurance against being hit by a first strike. Of course, one might say if there is that much trust, insurance could perhaps be bought much more cheaply. Perhaps some submarines would suffice.

So, we might consider the two cases : that there is trust and that there is no trust. If there is no trust - and obviously that is the present situation - then SDI will be seen as a protection against the second-strike capability of the opponent. That will not lead to a reduction in offensive arms but to intense efforts to circumvent the defense including increases in offensive weapons. We are also risking disharmony among the allies because of differences in the protection against Intercontinental Ballistic Missiles (ICBMs) and Tactical Ballistic Missiles (TBMs), and because of the neutralization of the nuclear shield for Western Europe and the ensuing necessity for increased reliance on conventional armaments which are more expensive and which are not provided for to a sufficient extent in the European budgets.

If trust could be created somehow then SDI could be rather limited and would not be destabilizing. Trust, in this context, would not mean that the two sides must love each other. They could still despise each other and even fight each other economically or ideologically. But they would have to aggree that military aggression is ruled out. They would have to trust each other at least in this respect; that would be sufficient. But, of course, a limitation of this kind, a limitation of trust only to the field of military aggression, would be difficult to achieve. It would certainly be helpful if there was co-operation in many fields so that the policy of non-aggression would become more credible than if you just had to say : Well, we hate each other but we won't attack each other.

The lack of trust, which is the reason for the problem with the consequences of SDI, is, of course, also responsible for the problems with arms control in general. As long as both sides suspect each other of just waiting for an opportunity for attack, at least in the long run, even if it's not to be expected tomorrow, there can be no decisive progress in arms control. Is there any hope for creating trust of this kind between democracies and the Soviet Union with its closed system of secrecy, lack of individual freedom and disregard for human rights? I think there may be some hope because the Soviet Union is not interested in suicide. It is also interested in international stability. This interest does not depend on internal reforms, although reforms might help for creating trust. But after 70 years of Soviet rule, the leaders of the Soviet Union have very little scope for reform. They cannot be very flexible, they can only make small steps towards more openness. We must look very carefully for even small signals. In our own interest we should respond to such signals and encourage steps in the right direction. That will be a slow process until the bulk of the system changes. The belief that their system must be seen as a threat to our system will probably survive for quite some time. But we must realize that our system is perhaps an even greater threat to their system by its greater attractiveness. Nevertheless, it will be to the advantage of both sides to avoid war. So we shall have to compete in other ways than by threatening each other in a mortal way.

What is the final outcome? Under conditions of no trust, SDI is likely to lead to even less trust and to an increased arms race with an enhanced risk of instabilities. Under conditions of "limited trust" (the two sides do not necessarily trust each other except in the mutual conviction that the opponent has excluded military aggression from his arsenal of options), a limited system defending second-strike capabilities would be stabilizing. But this trust has not yet been established. Without it, SDI will enhance suspicion. The establishment of trust, even "limited trust" will take time. Co-operation in as many fields as possible will help to establish it and, of course, so will the exchange of people.

One way to create more trust and to remove misunderstandings is to study the developments on the other side and particularly the debate on long-range goals.

From the discussion

Richard L. Garwin:
Klaus Gottstein, in your summary you questioned whether the SDI would be protection of a second-strike capability, or protection *against* a second-strikecapability. This point deserves additional emphasis.
Klaus Gottstein:
I said "against". As long as the two sides do not trust each other, SDI is seen by the other side as protection against its second-strike capability.
Richard L. Garwin:
Protection against a second-strike capability would be destabilizing.
Klaus Gottstein:
Yes. If SDI were seen as protection *of* a second-strike capability, that would be good. But that would require trust. Unfortunately, each side suspects that the other side is planning a first strike and that it is building up an ABM system to protect itself against the retaliatory strike. And that, of course, is the whole crux of the issue, because everybody thinks that SDI is not a protection against the first strike, but against the second strike.

Richard L. Garwin:

Finally, of course, if one has only ballistic missiles, then it would be very difficult to achieve a mutual judgement that defenses on both sides are only insurance against a first strike but have no effectiveness in protecting against a second strike, because the defenses are much better in protecting against a second strike at a time of one's own choosing. So, there ought to be, I think, some comment about that as a problem which must be overcome. If one has additional deterrent forces retained, such as the cruise missiles, there is no need for this insurance.

Klaus Gottstein:

I agree. Mistrust is the problem that must be overcome. I don't believe that technical fixes can do this. If the basic mistrust continues, then you will think that these cruise missiles and submarines are there to deter against a second strike. And SDI will be seen as an attempt to protect your population against a second strike. In short, any device will be seen as a trap, as a help for aggression, if you expect aggression. If you walk through Central Park at night, and you meet somebody who looks very suspicious and carries a knife, you don't assume he does this to protect himself. You are afraid that he will attack you. However, if it is a very kind old lady whom you know, then you will assume that she believes that this will give her some protection. Or, here is another example: Canada is not afraid of the nuclear weapons of the United States, although, of course, the United States could just as well wipe out all the major cities of Canada. Technically, this would be feasible, but the United States and Canada are friends and nobody expects that.

Robert S. Cooper:

Without commenting at all on targeting policy of the U.S., it certainly is clear that both the Soviet Union and the U.S. have the capability to strike each other's land-based weaponry. There is more or less capability, depending on what each side has provided themselves. Those weapons have been created, I imagine on both sides, out of a desire, by both the political and military structure, to limit damage in case of a nuclear war, because both sides believe that in case of nuclear war, some missiles would be released, and there would be others remaining, and as long as there were high accuracy warheads available to them, that you could limit further damage, by striking these forces' second strike.

Now, those highly accurate weapons are also useful in a first strike arrangement, but it seems to me that - whether you have defense or no defense - the first step in getting to this position of trust is, for the political and military authorities on both sides, to stop asking their military target planners to hold whatever forces remain after a first strike at risk by providing these high accuracy weapons to do so. Until that happens, there won't be any trust. Until those weapons are done away with, there can't be any trust between the two sides.

Klaus Gottstein:
It is the political stance which poses the threat, and not so much the existence of the weapons. The Swiss have a very strong army, they have a considerable bomber fleet, and they keep it ready. They could, at any time, destroy Milan, Freiburg, and other cities. Technically they would be able, and perhaps they are even targeting for training purposes. I once talked to a high-ranking Swiss officer, and I mentioned that the Swiss Army could, without too much difficulty, march to Milan. He laughed, and he said: "Well, we could actually march to Rome, and in our officers' exercises, we have done this as a war game; beating the Italians and marching to Rome." Probably the Italians know about this, but they are not very much upset.

Robert S. Cooper:
I should point out that if the Canadians had 25 holes in the ground with even intermediate range missiles in them, those missiles would probably be on somebody's target list. Trust is what you define it to be.

Richard L. Garwin:
I just want to support a very specific point in what Bob Cooper says - that a big problem is the accurate weapons which were purchased at considerable cost by the military and the civilian leadership to threaten the retaliatory force of the other side. Even if these are supposedly to be used in second strike mode, they have a better first-strike capability against the retaliatory force of the other side; and that engenders a kind of distrust which the ability to destroy the other side's cities does not.

Robert S. Cooper:
Which neither side intends to use.

Klaus Gottstein:
It is a vicious circle, actually. Because there is no trust, you take these precautions, and you do this targeting, and then, of course, this targeting confirms the suspicions of the other side, and there is even less trust afterwards.

14 Trust Creation by Discussion of Long-Range Goals [37]

At Munich, in 1985, we have begun a project in which, with the aid of experts on Soviet affairs, we intend to analyze the discussions going on inside the Soviet Union about the *long-range goals of Soviet Society*, both in internal development and in external politics. we hope that we shall be able to identify certain schools of thought, distinguishing between what is taught in school about the unavoidable demise of the capitalist system and what is actually influencing the actions of the Politbureau, or may be expected to do so some time to come.

I do not expect that the experts on Soviet affairs will come to complete agreement on the schools of thought which struggle inside the Soviet Union about the goals and methods of future policies, internal and external, on the hawks, the doves, and the various in-betweens, and their respective fictions and ideologies. We shall not try to determine who is right and who is wrong and who will win. But I hope it will be possible just to list these various schools without passing judgement. We shall then try, with the help of experts on Western political affairs, to determine, for each of these Soviet "options", adequate Western policies which would represent Western interests without leading to unstable, war-prone situations. Western reactions would have to be different, of course, under different assumptions about the long-term goals of Soviet policy. If, e.g., it were assumed that the expansion of the Soviet sphere of influence by all means, including military force, were the ultimate goal of Soviet policy, the adequate Western policies would have to be different than if the mere consolidation of the present sphere of

---

[37] From a lecture given at the Center for Science and International Affairs, Harvard University, Cambridge, Mass., 6 February 1985. Published in: Jahresbericht 1985, Forschungsstelle Gottstein in der Max-Planck-Gesellschaft, München 1986. We quote this text here as an example of a scientific project for the identification of possibilities for creating "mutual trust" in a situation of hostile confrontation such as existed during the Cold War. The first part of the lecture is reproduced in the section *Possibilities for creating mutual trust* of chapter 12. The project, as described in chapter 14, was carried out and resulted in a series of conferences, with participation of scientists from the United States, the Soviet Union and Eastern and Western Europe, until the end of the Cold War. Some of the results are given in the following chapters.

influence, and peaceful competition with the West in the economic and social fields, were assumed to be Soviet goals. The latter goal, or option, peaceful competition, could possibly be nurtured by a Soviet belief, as expressed by Khrushchev, that socialism of the Soviet type will prove its superiority in the long run, and that Western capitalism will decay by itself.

Listing such assumptions we shall get a table of possible Soviet intentions and of appropriate Western reactions. The next step will then be to show this table to influential Soviet scholars and scientists. Our Soviet counterparts will then get a better feeling for the impression which Soviet political moves create in the West, and for the Western reactions which might be expected to result. Actual political situations might be linked, or compared, to the options and counter-options in the table, and their effects on stability against Sarajevo effects might be discussed. This way the likelihood for dangerous miscalculations may be reduced to some extent, in that we contribute somewhat towards mutual understanding of the very different ways of thinking of the two sides. We shall also try to persuade our Soviet colleagues to perform a similar appraisal of Western intentions and corresponding Soviet reactions. For them, of course, this will be much more difficult because there is already an official appraisal, an official party line which cannot openly be contradicted. Nevertheless, even if we do not get a reply we can give this exercise of listing Soviet assumptions about Western intentions, and about likely Soviet reactions, to the official party machinery through our Soviet colleagues. This alone may have an educating effect within that machinery. And education is always helpful. If we are lucky we may be able to compare results, their table with our table, and determine policy options which are particularly dangerous because of the instabilities to which they may lead. One can then discuss, each side within its system, and between the systems, how best to avoid these instabilities.

The main purpose of this project is to find out about possibilities to create trust between the two systems: trust not necessarily about their intentions but about their methods. It is like confirming the rules of a boxing match. Each participant knows that his opponent wants to knock him out. But he also knows that he wouldn't hit below the belt. We must try to find similar rules for the East-West confrontation, as long as it lasts. The equivalent of hitting below the belt in this example would be measures leading to instabilities with respect to nuclear war.

15 Western Perceptions of Soviet Goals [38]

An educating example from the beginning of the end of the Cold War

The relations between East and West are in a state of transformation and re-definition. It is not clear yet which form they will take in the years to come. One basic fact, however, remains unchanged from years past: Even though we do not know the future, even though politicians in East and West often do not, and cannot, know the long-range plans, intentions, and imaginations of their opposite numbers, they do have some perceptions, consciously or subconsciously, of potential developments for which their own side will have to be prepared. The atmosphere between NATO and the Warsaw Pact, certainly, is now more relaxed than ever. However, we cannot be sure that a relapse into the Cold War will not happen. But even if such a relapse will not occur, the degree to which East-West co-operation will become possible in arms reduction and in the tackling of environmental and other global problems will certainly depend on the mutual perceptions of the long-range goals of the other side.

At this Research Unit of the Max Planck Society we decided, in 1984, to investigate various Western views on the potential future developments in the U.S.S.R., as well as Soviet views on the future developments in the West, as far as they influence long-range planning. The method chosen was to organize conferences in which Western observers present their perceptions of Soviet developments, and Eastern observers their perceptions of the developments in Western countries. The arguments given to support these perceptions are then to be discussed among the participants. [39]

As mentioned above, we have selected this topic of the perceptions, in East and West, of the long-range goals of the other side because it seems obvious that these perceptions are the main obstacle to arms reduction and closer co-operation between East and West. So far, on both sides, politicians seemed to

---

[38] From Preface, Introductory Remarks, and Round Table Discussion in:
K. Gottstein (Editor), *Western Perceptions of Soviet Goals: Is Trust Possible?*
Campus Verlag/Westview Press, Frankfurt am Main/Boulder, Colorado, 1989.
[39] See chapter 14.

consider it prudent to base their long-range planning on worst-case scenarios with respect to the intentions, or "goals", of the other side. The "goals" as such may not even exist in reality - at least it will be very difficult to identify them in the heads of leaders - but the perceptions of these goals do exist in the worst-case scenarios prepared by both sides. They are real because they determine policies. The worst-case scenarios about what the other side might be planning to do are often based on experiences of the past which are no longer valid, on wrong interpretations of observed developments, incomplete or faulty information and on misperceptions and misunderstandings of various kinds. What are their sources? It is important to analyze the origin of these views. On what kind of education or indoctrination, on which ideas, rational arguments, social structures, historical memories, biographical experiences, ideological biasses, enemy images are they based? To what extent will it be possible to identify some of these perceptions as clearly due to misunderstandings? Which of them may be considered as "normative" or "strategic" in the sense that they are intentionally and/or artificially upheld in order to influence one's own public or that of the other side?

How does each side perceive itself? How does each side believe it is perceived by the other side? In particular: What do we expect from each other? What could be done, in the long run, to avoid crises and improve co-operation under existing perceptions? In what respect are the perceptions changing? If it is co-operation that is required for solving global problems, what type of changes are required in the mutual perceptions of "Grand Strategies"? Which measures would have to be taken by either side if existing perceptions were to be changed intentionally in a direction representing greater trust? (A well-known example from the military field: The Soviet Union claims to have a defensive doctrine but, until very recently, considered the capacity for vigorous counter-attack as the best defense. To Western observers this capacity looks like readiness for attack. Soviet analysts consider this conclusion to be a misperception. A measure helpful for changing this alleged misperception would be a restructuring of Soviet forces so that fears of a sudden attack are allayed. Similar examples can be found in the fields of foreign policy, economy, ideology, etc. I am sure that Soviet observers could give examples where Western behaviour seems to contradict Western policy declarations, and where Western re-arrangement measures would help to allay Soviet security concerns.)

If some changes of basic beliefs are difficult or impossible to effect in the near future, to what extent can we live, and cooperate, under the "old" perceptions? Are there specific recommendations that could be made under different perceptions and assumptions? Are there areas of consensus, and which are the arguments leading to dissent?

We began by convening Western scholars, diplomats, representatives of industry and military experts to consider a problem which was, and still is, a major subject of controversy in East-West relations as well as in inter-alliance debate: the impact of the SDI project on strategic, crisis and arms control stability.[40] One main result of that conference, held near Munich in 1986, was: the judgement whether SDI has a positive or negative effect on stability depends on the mutual perceptions of the long-range intentions of the other side.[41] Our next step was an International Workshop on Western perceptions of long-range goals of Soviet policy, held at Ringberg Castle, the Conference Centre of the Max Planck Society, in November of 1987. It was attended by 27 scientists, scholars, and diplomats from nine countries.

In this project the term "goals" is not used in the sense of conscious objectives but in the sense of perceived driving forces behind Soviet political developments, inasmuch as these perceptions shape Western reactions to Soviet moves as well as Western initiatives towards the U.S.S.R. The spectrum of Western perceptions of such Soviet "goals" is, of course, an almost continuous one, and it may seem arbitrary to pick a limited number from the host of views expressed in the literature and in public statements. It was tried to limit the discussion to those which are of political significance, or may become influential in the future.

Mutual perceptions of long-range political goals as an obstacle to trust creation

The idea that it is really the perceptions of the intentions and goals of the other side which matter for our reactions, and not the intentions and goals as such, is not new. General and Professor Count Baudissin, one of the fathers of the Armed Forces of the Federal Republic of Germany, used to say that it is not so important whether we are secure but whether we feel secure. He also

---

[40] See chapter 13.
[41] K. Gottstein (ed.): *SDI and Stability. The Role of Assumptions and Perceptions.* Nomos Verlagsgesellschaft, Baden-Baden, 1988 (See also Chapter 13).

said that states that feel insecure become incalculable in their policies. And he noticed that our own security depends on the feeling of security of our opponents. It seems obvious that trust can only be developed if we believe that we understand the intentions of the other side and identify them as being peaceful.

President Richard von Weizsäcker, in a speech in London last year, said: "True, the two blocs mistrust each other because they are armed. But it is also true that they are armed because they mistrust each other." President Reagan put it differently when he visited West Berlin. "No", he said, "East and West do not mistrust each other because we are armed. Perhaps we are armed because we mistrust each other." In this case, I think, President Reagan came nearer to the truth: It is not the arms as such but the perceived or suspected intentions behind the arms which create mistrust and misgivings. Britain and France are heavily armed, and so are France and Germany. But they do not mistrust each other with regard to the intended use of these arms. I think President von Weizsäcker would agree.

In this context, Richard von Weizsäcker, on several occasions, made very wise remarks. When in Moscow he spoke about the importance of cultural exchange. The political and social systems in East and West are very different, he said, and that results in a lack of knowledge about each other and about the motivations of each other. By co-operation, e.g. in cultural exchange, one could remove "the unjustified part" of mutual mistrust, the President declared. Weizsäcker thus distinguishes between a justified and an unjustified part of mistrust. That is a very good example of perception as well as of self-perception. Weizsäcker also told the Russians that both sides should refrain from irritating each other, and he reminded them that coming to an agreement with each other does not mean approval of the other system.

I have given these examples in order to illustrate the relation between perception of intentions, and trust or mistrust. But these examples already showed that there are different categories of perceptions. Apart from the perception of one side by the other side, there is self-perception (how does each side see itself?), imagined self-perception (how does each side think that the other side sees itself?), and imagined perception of the other side (how does each side believe that the other side sees one's own side. All of these categories can be used in a strategic way (by *pretending* to perceive ...) or in a normative way (by *wanting* to perceive myself or others in a particular role). So we have at least 12 different kinds of perceptions.[42]

Moreover, there is the complication that nations, let alone blocs, are not uniform. Different individuals in one nation may have different perceptions, and we are confronted with the task of guessing, or perceiving, which of these alternative perceptions will surface and determine policy some time in the future. In spite of all this complexity, no policy-maker can avoid having perceptions of his own side, and of his opponents, and he had better find out what they are and how they took hold of him. He also should find out what the perceptions of his opponents are, the true perceptions as well as the strategic and the normative ones. Unless policy-makers and their advisers learn to do that, their policies will continue to suffer from misjudgements and will lead into all kinds of traps.

Of course, we must realize that whatever we find out in the West about Soviet perceptions and Soviet self-perceptions, once we articulate our findings, once we formulate them, our opinions become Western perceptions. They may be Western perceptions of Soviet perceptions of Western perceptions. But that is reality. We can't get around it. And the Soviets are, of course, in a reciprocal situation with respect to their perceptions of our perceptions. There is no absolute truth in these mutual perceptions. The only way out is to sit down with our Soviet counterparts, compare results, and try to avoid self-perpetuating misunderstandings and undesirable consequences in which no side is interested. That does not even require trust. It just requires a certain minimum of common interests.

Let me say a few words about the second part of the title of our Workshop, i.e. about the so-called long-range goals of Soviet policy. As you know, this Workshop is about "Western perceptions of long-range goals of Soviet policy". Some people have objected that there may be no long-range Soviet goals, just short-run improvizations in response to shifting circumstances. That may be true but it does not really matter as long as there are Western

---

[42] Michael Richter, Meaning and Perception Patterns: A Preliminary Logico-Semantic Analysis. In: K.Gottstein (Ed.), *Western Perceptions of Soviet Goals: Is Trust Possible?* Campus Verlag/Westview Press, Frankfurt am Main/Boulder, Colorado 1989.

Michael Richter, Political Understanding, Perspectivism and Dialogue Structure. In: K.Gottstein (Ed.), *Mutual Perceptions of Long-Range Goals*. Campus Verlag/Westview Press, Frankfurt am Main/Boulder, Colorado 1991.

Michael Richter, The Perception Method for Analysing Political Conflicts. In: K.Gottstein (Ed.), *Tomorrow's Europe*. Campus Verlag, Frankfurt am Main 1995.

perceptions of Soviet long-range goals. If Western policies, e.g., were determined by a perception that the long-range goal of Soviet policy will continue to be the spreading of the Soviet type of socialism across the globe, even if such a goal did not exist any longer, the Soviet Union would be forced to react to Western measures trying to contain the Soviet Union. If, on the other hand, Soviet perceptions were that the West, in the long run, would try to conquer or subjugate the U.S.S.R., even if that were an erroneous assessment, the West would have to cope with Soviet policies trying to forestall this subjugation.

So again, in this kind of game, the perceptions matter at least as much as the facts.

There is another consideration which makes these long-range goals important even though they may not exist as concrete objectives. Long-range goals can be subconscious. They are the result of education, of cultural identity, of ideology, they are in the back of the minds of policy-makers when they make their improvizations and small steps responding to the requirements of the day. The direction of these small steps may be affected by the ideological notion, in the back of the minds of Soviet leaders, about the long-range role of the Soviet Union in world society. Even if there is no planned objective, the small steps will not add up to a random walk in this case but to an overall move in a certain direction. It will be worthwhile to try to analyze this direction in order to be prepared for possible developments. At least, and here the concept of perception enters again, we should be aware of the fact that the other side will be influenced by what they perceive as the direction of our steps, and vice versa.

The reforms announced by Gorbachev - perestroika and glasnost - have caused an avalanche of speculations in the West about the future of the Soviet Union, the chances of success for Gorbachev, and what that would mean for Soviet policies, and for East-West relations. Our project was started before Gorbachev was appointed Secretary-General. We started from the observation that the mutual perceptions of long-range goals are the main obstacle to agreements on arms control and on co-operation in other fields. The reduction of these obstacles - their removal, if possible - was the main motivation for our project, and it still is. The appearance of Gorbachev certainly affected some of the Western perceptions of Soviet long-range goals. Therefore, we have to study most carefully the impact of Gorbachev's moves. But Gorbachev did not change all of the Western perceptions of the Soviet Union. The skeptic and pessimistic perceptions are still alive in many circles and will have to be taken into account in political assessments. We have to register *all* of them, as far as they are significant. We do not depend on optimistic expectations that the Soviet Union is going to move now not only towards modernization, but also towards Western norms of democracy and to a reduction of its external pressure so that containment and deterrence will soon become obsolete. It would be nice if that were true, but so far it is only the perception of a minority in the West. Western policies are still determined by other perceptions. We shall try to list *all* perceptions. Henry Kissinger, just sees the East-West conflict as a normal competition for supremacy of a type which history has seen many times in the past and which has nothing to do with Western values or their absence; after all, Machiavelli was a product of the West. If the Tsar had his troops on the river Elbe, instead of the Soviet Union, and were trying to expand his influence even further, the situation would not be much different from what it is now, Kissinger once said.[43]

In any case, opponents must learn to understand each other if peace is to be preserved. Technical measures of arms control and confidence-building are not enough as long as we consider the other side to be our enemies. As Dr. Rita Rogers points out,[44] people are uncomfortable with the uncertainty of not knowing the motivational attributions of their counterparts. If they do not know these motivations, they invent them. Lack of knowledge creates strong incentives to prepare for the worst. Unfamiliar problems are often erroneously discussed in terms of the familiar. Therefore, it is so important to work for mutual understanding of goals and motivations, and to avoid tension and hostility which create military needs which in turn enhance tension, mistrust and hostility.

The picture just given of the role of mutual perceptions, and in particular of Western perceptions of the discussion going on inside the U.S.S.R. on the goals of Soviet development, was certainly an oversimplified one. Our Western perceptions of the role of partisanship in Soviet ideology, of the lack of personal freedom were not mentioned, nor the dialectic method applied to politics which allowed the Soviet Union right from the beginning of the

---

[43] Henry Kissinger, private conversation with the author, Cambridge, Mass., 26 August, 1963.

[44] Rita Rogers, Fear of the Soviet Union: Individual and Cultural Reflections. In: K.Gottstein (Ed.), *Western Perceptions of Soviet Goals: Is Trust Possible?* Campus Verlag/Westview Press, Frankfurt am Main/Boulder, Colorado 1989.

1920s to follow a dual course and to cooperate with foreign governments which they tried to overthrow at the same time. As early as 1927, at the World Economy Conference in Geneva, the Soviet delegation advocated the peaceful coexistence of the two economic systems. Simultaneously, the policies of COMINTERN were directed against the Social Democratic Parties in Europe, and for revolution in the capitalist countries. We have a different situation today, COMINTERN no longer exists, but the dialectic method is probably not forgotten, and dialectic discrepancies between foreign policy, military policy and economic policy cannot necessarily be excluded without further inspection. It is important for the problem of creating mutual trust because what looks like dialectics to Soviet-trained observers looks like lying to Western observers. At least that is what George Kennan noted (DIE ZEIT, 17.02.84). Erhard Eppler, the Chairman of the Committee on Basic Values of the Social Democratic Party of Germany, criticized Kurt Hager, the Chief Ideologist of the Communist Party of the German Democratic Republic, for continuing to use the term "enemy" for the capitalist system in a 1987 newspaper article, and for denying the capability for reform and for peacefulness to that system. He did this right after representatives of the two parties had agreed that peaceful co-existence and even co-operation between the two systems is possible. This is just another example of dialectics, and it is surprising that Erhard Eppler is upset. It would indeed be sensational if this kind of dialectic double-talk were given up. From a marxist-leninist point of view there is nothing wrong with it.

From the discussion

Arnold Buchholz:
It would be useful to arrange in some kind of matrix the long-range trends of the Soviet Union. Along one axis, the long-range trends of the Soviet Union, and along the other axis, the perceptions of these developments, so that it would be possible to have these different possibilities to make classifying possible.

Klaus Gottstein:
I think what Dr. Buchholz has in mind is to investigate separately the various fields of Soviet development, under different perceptions that is the cultural field, the foreign relations field, the ideology, the economic field, the security field, and so on, and then discuss each of these fields separately as to its potential development. This is, of course, an even greater task than the

already very large task which I have set for our group, but partly, I think, it coincides with what we are trying to do. We have had - or are going to have - separate lectures on the cultural field, on economic aspects and on security aspects, on education, and on science and technology policy. However, what I had in mind was more modest. I only wanted to look at those aspects in the cultural field, in the economic field, in the science and technology field and so on, from which one can learn something about the long-range goals of Soviet society , inasfar as East-West relations are affected by them. So we might be able to drop some of the spheres which are only of internal importantance for the Soviet Union without affecting East-West relations very much. For instance, in the sphere of sports, which emphasis the Soviet Union might place, in the long run, upon light athletics as compared to some other fields would not affect East-West relations very strongly although it might be very important for the sports policy of the Soviet Union.

The main motivation in this project is really arms control, starting from the observation that the principal obstacle to progress in arms control seems to be the mistrust concerning the long-range political intentions of the other side. Even if they are not yet formulated, only existing in the back of the minds of the leaders, they have their effect. These are the areas on which we should concentrate, namely those which are most likely to affect Soviet mistrust against the West and Western mistrust with regard to the Soviet Union. So, if we apply this selection criterion, some of the other fields might have a lower priority, although I would admit that probably every field has some kind of impact on East-West relations, but probably some more than others. And therefore, since we cannot do everything anyway we should concentrate first on those things which probably have the greater impact, and these are the fields of security policy, foreign relations and economic policy.

We do not yet know how far-reaching glasnost will become, so perhaps sometimes we shall have to read between the lines of official statements regarding long-range political goals. We may, for instance, also learn by studying Soviet school books. We won't be interested - at least as far as I am concerned - in Soviet education policy as such, in all fields, but only what children are taught in school-textbooks about the West and specifically, what they are taught about the need to defend the Soviet Union against a future attack by the West. Such texts and similar statements might give a clue of Soviet visions of the long-range future. I would be interested in that part of educational policy. This will also apply to other fields. If theatrical productions are encouraged to show the West as the enemy, for instance, or show the West as co-operative, then I would become interested, but not in theatre as such. Perhaps that explains how I would suggest to set the priorities.

Eberhard Schulz:
There are fields of Soviet domestic development which have a strong im-pact on Western perceptions, such as the treatment of dissidents, the nationalities problem, the treatment of Soviet and Russian history, the publishing of literature. It seems to me that we should include these typical aspects, as they have an impact on Western perceptions, although they are not immediately related to foreign relations.

Klaus Gottstein:
The treatment of dissidents and nationalities would only interest me inasfar I can learn something about long-range Soviet goals from it. These topics are very important for the internal development of the Soviet Union but they have aspects which may not be that important for learning something about long-range goals.

I would not be happy if we only produced a research volume which would then be shelved in the library and nobody would ever look at it again. I think our results should really influence the political process. What we learn should be available to the decision-makers. Of course, a research volume is a basis for this. If we just talk then it will be forgotten, so there must be something on paper.

Arnold Buchholz:
If we look at the phenomenon of the so-called perestroika, we can see that its development is very unequal in the diverse components of the system. In the field of economics, perestroika is making very slow progress, whereas on the historical sector, very much is under way. Then there is the field of literature, in which we find a very progressive development, and attempts are being undertaken to apply perestroika to ideology. New approaches are also being observed in the area of human rights. ... We know that the Soviet concept of coexistence does not mean ideological coexistence. Ten years ago Giscard d'Estaing, the French president, said in Moscow that it would be useful to have spiritual coexistence. And the Russians said, "No, that's impossible". But now there are new possibilities to be found in specific fields; there is a broader basis for spiritual co-existence than there was ten years ago. First of

all, there are the problems of mankind's survival and some common human values. At the same time, the weakest points of the Soviet spiritual and cultural life - one can say of the whole system - are to be found in the fundamental problems of human existence, like death or the meaning of life. This has now become apparent to Soviet philosophers. I think it would be possible to discuss in a reasonable way with Soviet specialists how to find a broader consensus in some fields of spiritual life so that it can serve as a basis for making the existing political and social conflicts more bearable.

Klaus Gottstein:
The final goal of the project is to make a contribution towards the removal of misunderstandings and of mutual fears based on misperceptions of the other side. ... Through feedback, misperceptions tend to strengthen each other. During the last stage of the project, Soviet and Western representatives are supposed to exchange their perceptions, and discuss the observations and arguments which lead to these perceptions. My hunch is that this process, as such, will make a considerable contribution to better mutual understanding and to the removal of misunderstandings. Often, some measures which one side takes with the intention of making its own side more secure look threatening to the other side. However, if the arguments are discussed one might come to the conclusion that with a small change the security requirements can still be fulfilled without appearing threatening to the other side. That is the final objective. There may be other by-products, but I think this is a worthwhile task in itself. One by-product will be that additional channels of communication will be established. There will be counterparts who can talk about these things with each other. ... When I talk about long-range goals I mean those goals, apart from many other goals, which look threatening to the other side. If the electrification of Siberia is mentioned as a long-range goal, I'm not interested within the context of this project. I might be interested in some other context, but not in this one. If, however, the spreading of Soviet-type socialism all over the globe is perceived as a long-range goal in the back of the minds of Soviet leaders, then this would be something which Western people consider as being a threatening goal. And, therefore, this should be addressed. So, we shouldn't discuss all kinds of goals - that would be much too wide a field.

16 Mutual Perceptions of Long-Range Goals[45]

Can the United States and the Soviet Union co-operate permanently?

From the preface

It is still not clear what the final outcome of the recent historical changes in the political landscape of Europe will be. Will there be a new, stable equilibrium, with a prospect of continued peaceful development? Or will the local and regional instabilities, already visible in some parts of Eastern Europe and of the Soviet Union, also affect international relations, particularly those between the United States and the Soviet Union, between NATO and the Warsaw Pact, and between the major European nations?

Whatever the path of history will be, it is clear that it will be deeply influenced by the perceptions which the acting elites hold of each other's long-range goals in the political re-orientation of international relations. The "goals" as such may not even exist in reality - at least it will be very difficult to identify them in the heads of leaders - but the perceptions of these goals do exist in the scenarios prepared by both sides. They are real because they determine policies. Will these perceptions allow mutual trust? Without mutual trust it will hardly be possible to deal satisfactorily with the tasks that will have to be carried out for the creation of a new, stable and acceptable European order.

At the time of the Cold War the course of events was essentially determined by the United States and the Soviet Union, with the European nations trying more or less successfully to have their interests taken adequately into account. Therefore it seemed reasonable to study next the mutual perceptions that can be identified among the political leadership and the public of the two leading powers. So we continued the project by convening a "Conference on the Mutual Perceptions of Long-Range Goals in the East-West Conflict" to which Soviet experts on the United States and

---

[45]From: K. Gottstein (Editor), Mutual Perceptions of Long-Range Goals (Proceedings of a Workshop held in May/June 1989), Preface, Introductory Remarks, Introduction to Panel Discussion, and Concluding Remarks. Campus Verlag/Westview Press, Frankfurt am Main, Boulder, Colorado, 1991.

Western Europe were invited as well as U.S. and German Sovietologists.

Nine Soviet scholars from four institutes of the Academy of Sciences of the U.S.S.R. (the Institutes for World Economy and International Relations, for Europe, for U.S.A. and Canada, and for the Economy of the World Socialist System), ten scholars from the United States, specializing in Soviet affairs and in the role of perceptions in politics, and fifteen German experts attended the meeting which took place at the Evangelische Akademie Tutzing, near Munich, from May 28 to June 1, 1989.

The spectrum of views on both sides regarding the opposite side may be divided with some simplification into the three schools of the "essentialists", the "mechanists" and the "interactionists". The essentialists perceive the other side as an essentially hostile power which can only be dealt with by confrontation; the mechanists view it as governed by geopolitical considerations, reacting in opportunistic ways, so that negotiations should best be undertaken from a position of strength; the interactionists, finally, believe in change by rapprochement, in the existence of possibilities for peaceful competition and for constructive co-operation.

The present openness of the Soviet Union toward the West was described as a structural phenomenon and attributable to changes in Soviet perceptions of capitalism, East-West relations and of the role of military power in world politics.

As the most significant of these changes in perceptions the following three were named:

- The erosion of the traditional image of an irreconcilable struggle between capitalism and socialism, in parallel with a spreading conviction of the interdependence of one world, the problems of which can only be mastered by joint efforts of mankind as a whole. Whereas only a few years ago spiritual co-existence with the West was considered to be impossible , the common belief in human values is now deemed to be the basis for co-operation.

- The observation that capitalism, contrary to Marxist predictions, did not collapse. In developing "New Socialism" the practical experiences of capitalism should be studied and used. Whereas in the past it were the weaknesses of capitalism that were of interest, it is now the task to learn from its strengths.

- The understanding that the political usefulness of military power is very limited, military expenditures leading, economically, to "red figures".

Western participants observed that the Soviet Union seemed to undermine its ideology without abandoning it. If capitalism isn't doomed, what is left of Marxism? Gorbachev constantly invokes Lenin. But it was Lenin who abolished freedom of the press and democracy.

Soviet participants replied that the type of capitalism which existed at the time of Marx did indeed perish. The West developed in a positive sense. Today the Soviet Union, by undergoing changes herself, would like to support this development.

Ideological questions did not exist any longer, Soviet participants remarked, except for conservatives. Lenin himself underwent changes during his lifetime. One has to distinguish the Lenin of before 1917, the Lenin of "War Communism" (1917-1920), and the Lenin of 1921 and after. Lenin's policies in his last period, with the new economic policy, were similar to perestroika. Unfortunately, some forces in the West, by their utterances, strengthened the old confrontational thinking in the Soviet Union.

Détente, it was pointed out, had different connotations in the United States and in the U.S.S.R.: The United States expected the U.S.S.R. to follow certain rules of good behaviour. The U.S.S.R. expected from the United States to be accepted as being of equal status.

With respect to the goals of Western politics the famous 1947 article by George Kennan was quoted: In view of communist expectations of an impending self-destruction of the capitalist world it should be the goal of long-range policy in the United States to show the capability for dealing effectively with domestic problems as well as for shouldering successfully the responsibility as a world power, while at the same time, with all the mental vitality of the United States, taking an appropriate place in the field of the ideological currents of our time.

In the prevailing atmosphere of the conference it was not surprising that proposals for a future policy of peaceful coexistence were mostly formulated from an interactionist point of view: areas of congruent interests should be identified, military structures should be changed in the direction of non-provocative defense, economic co-operation should be intensified, contacts in as many fields as possible should be cultivated, and a political code of conduct should be agreed upon. But it was also made clear that mechanist and essentialist attitudes were still alive. There is no guarantee that their influence on policy-making could not increase again when circumstances changed. There is much concern in many circles about European stability and global equilibrium. It therefore remains important to understand the sources which feed potential mistrust, misunderstanding and ill-feeling. The interactionists as such need not be convinced of the value of honest co-operation, but the mechanists and essentialists do. If interactionism is to survive and determine the future course of interna-tional relations, it is necessary to cope with essentialism and mechanism. This can only be done if the roots of their enemy images and their changing sources of support are analysed, understood and constantly monitored. Of course, with Western enthusiasm for perestroika and glasnost and with the new Soviet and East-European expressions of belief in market economy and democracy there is also a new risk arising from exaggerated friend-images. Expectations that are unrealistic and cannot be fulfilled may lead to disillusionment, public frustration, disappointment and, as a consequence, to political instabilities.

From the introductory remarks:

I am sometimes asked why I talk so much about perceptions rather than about observable facts. I think the best reply to that question was given by the Greek philosopher Epictetus who lived from 55 to 135 A.D. and who is reported to have said: "It is not the things themselves that make us happy or unhappy, but our perceptions of the things." It seems obvious that what makes people in the United States happy or unhappy about the Soviet Union and people in the Soviet Union happy or unhappy about the United States is also strongly influenced by the *perceptions* of the facts. And when people start talking about the long-range plans of the other side, when they say "Communists will always remain Communists" or "Capitalists will always remain Capitalists" and then continue "These people on the other side try to outsmart us, they only pretend to be cooperative, their long-range plan is to bury us", then it is particularly clear that people are not talking about realities, but about their perceptions of what the realities seem to be. And these perceptions can be - and they are - a formidable obstacle not only to disarmament but also to honest co-operation. They lead to mistrust. As psychiatrists know, deeply seated mistrust cannot even be removed by hard facts showing the opposite of what is actually feared. If mistrust reigns, even any friendly measure will be interpreted as a trap. As long as mistrust about long-range intentions is not removed, any measure, say of troop reduction or arms withdrawal or declaration of a defensive doctrine, will be seen as a trick to strengthen one's own side while lulling the other side into defenselessness, just a design of the sort which Hitler used when he built the defensive line of the so-called West Wall before World War II and declared a defensive doctrine for Germany. Shortly afterwards he attacked across that "defensive line".

The only way to distinguish removable mistrust from non-removable, that is justified, mistrust, and to remove removable mistrust is, as psychologists tell us, to try to put oneself into the shoes of the other side and to understand the arguments supporting the perceptions of the other side, and the history which led to these arguments and perceptions.

It might be said: "Why go to all this trouble? Can't we just wait and see and let history do the job?" Well, we should be careful. History is done by people, and people can be incalculable. At present there exists an opportunity for discussions like this one. Opportunities must be seized. They can go away. We don't know how long they will last. There have been relapses before from co-operation or detente into Cold War. NATO and the Warsaw Pact have existed for 40 years and peace in Europe has lasted about that long. But some historians remind us that alliances in history usually last for about 40 years: The Holy Alliance lasted from 1815 to the Crimean War of the mid-1850's, and the Austro-German Alliance formed by Bismarck to secure peace in Europe after the Franco-German War lasted four decades also. So we should learn from history and not take for granted the continuity of alliances and the perpetuance of peace. As long as military planners on both sides in their worst-case scenarios find it necessary to deter aggression by the preparation for nuclear warfare we are in a very dangerous situation even if we might be inclined to believe that this is the safest or perhaps the only way to preserve peace. But this very belief which makes us accept high risks for the preservation of peace is based on mistrust. This mistrust is based on certain perceptions. Perhaps if we succeed in changing these perceptions, on both sides, of potential long-range goals and intentions of the other side, we can reduce the risk.

We shall talk here about the perception of Soviet goals by the United States and of United States goals by the Soviet Union and of the self-perceptions of both sides. But, perceptions of the other side are strongly influenced by the imagined self-perception and by the imagined perception by the other side. For example, the way the Soviet Union is perceived by the United States is influenced by the way Americans believe the Soviets perceive themselves and by the way Americans believe the Soviets see the United States. But that is not all. When I hear somebody telling me about his perceptions of myself and of himself, then I cannot avoid asking myself what his motives are in telling me these stories. Is he telling me the whole story or is he suppressing some aspects of the whole story which he does not like? Particulary when self-perceptions and perceptions of the other side are proclaimed by politicians, we have to ask: Whom does he want to convince? Are the perceptions meant to be "normative", i.e. do they describe a situation which he would like to see and in which his fellow citizens are supposed to believe although he does not honestly see it himself? Or are they "strategic", i.e., are his adversaries to be influenced by the pronouncement of these perceptions? Michael Richter has categorized these different kinds of perceptions.[46]

There is also the interesting question of overlap. To what extent does my perception of myself and of my adversary agree with the perception which my adversary holds of myself and of himself? And what can be derived from stating such an agreement or partial agreement? In other words: For each perception there is a self-perception by the other side, and for each self-perception there is a perception by the other side. Moreover, for each perception and each self-perception there exists an imagined version by the other side. Now, I don't expect really that we can keep all these things apart. Very often we won't know, e.g., whether a perception is strategic, normative or true. But I think it is useful to keep these differences in mind, in principle. If we want to understand each other, we must try to accomplish the imagination mentioned above, and we must compare notes with the opposite side on the contents of our perceptions. We must explain why we perceive some of the statements by representatives of the other side as honest and

---

[46] Michael Richter, Meaning and Perception Patterns: A Preliminary Logico-Semantic Analysis. In: K.Gottstein (Ed.), Western Perceptions of Soviet Goals: Is Trust Possible? Campus Verlag/Westview Press, Frankfurt am Main/Boulder, Colorado 1989.

Michael Richter, Political Understanding, Perspectivism and Dialogue Structure. In: K.Gottstein (Ed.), Mutual Perceptions of Long-Range Goals. Campus Verlag/Westview Press, Frankfurt am Main/Boulder, Colorado 1991.

Michael Richter, The Perception Method for Analysing Political Conflicts. In: K.Gottstein (Ed.), Tomorrow's Europe. Campus Verlag, Frankfurt am Main 1995.

genuine, and some as strategic and intentionally misleading, and some as normative or as wishful thinking.

In our special case of East-West relations it will be interesting to see to what extent there will be consensus between the two sides, i.e. to what extent the perceptions of one side overlap with the self-perceptions of the other side, or to what extent the self-perceptions overlap with the imagined self-perceptions by the other side. It seems that there cannot be trust between the two sides unless there is a certain overlap of this kind. If all perceptions were seen as misperceptions by the other side, then there cannot be trust. Some of the psychological reasons for misperceptions are given by Professor Rita Rogers.[47]

The following is reproduced here as an example of how a meeting should be structured that is to clarify the mutual perceptions of two opposing sides, and remove misunderstandings. It is also taken from the introductory remarks of the author to the conference "On the Mutual Perceptions of Long-Range Goals in the East-West Conflict" in the early summer of 1989:

The structure of this conference is that we have two speakers from the Soviet Union and two speakers from the United States giving their perceptions of the other side, including the imagined self-perception of the other side, and two speakers from both sides giving their self-perceptions and the imagined perceptions. At least that was the intention. It may be difficult to carry this Programme out, in some cases. We will see what we will get when we hear the presentations. Each of these four papers will be commented upon by a speaker from the other side. I am very grateful to our colleagues Sergej Plekhanov, Raymond Garthoff, Vladimir Baranovskij and John Van Oudenaren for having accepted the task of writing the four main papers, to Professor Richard Pipes and Dr. Vladimir Babak, Dr. Victor Shein and Dr. Richard Herrmann for being willing to offer their commentaries. Of course, the other participants will have their chance to make their remarks. I hope that at the end we shall come to some sort of consensus or at least to an overlap of opinions as to what the various schools of thought are which

---

[47] Rita Rogers, The Kaleidoscope of International Decision-Making: The Human Factors in Crisis Management. In: K.Gottstein (Ed.), Mutual Perceptions of Long-Range Goals. Campus Verlag/Westview Press, Frankfurt am Main/Boulder, Colorado 1991.

represent the spectrum of perceptions on each side and what the policies are that follow, or should follow, from these different perceptions. I hope we shall also discuss the arguments which are used to support each of these perceptions. Are all of these arguments watertight or do we have to change some of the perceptions because we find that the arguments leading up to them are *not* watertight? I hope we shall also be able to determine what each perception means for the prospect of peaceful development, to what extent co-operation will be helpful, and to what extent we should restrain ourselves from giving advice to the other side, letting each side develop in its own way, perhaps in a convergent way, as, surprisingly, some of our Soviet colleagues have indicated. Are there any specific recommendations we should make at the end of this conference for dealing with each other in order to remove perceptions that could become dangerous, and in order to foster perceptions that lead to peaceful and mutually advantageous co-operation?

Introduction to the panel discussion

What do we expect from each other in the West and in the East? What could be done, in the long run, to avoid crises and improve co-operation under existing perceptions? We cannot expect to change the perceptions so easily, as the psychologists among us and those others who have read a little bit about psychology have already pointed out. So we must live with the existing perceptions and see what we can do under them. But, of course, we can also ask: In what respect are the relative weights of these perceptions changing? It is true that a person will not easily change his perceptions, but new persons are coming, old persons are dying or retiring. So the relative weights of perceptions do change.

If it turns out that co-operation is required for solving the global problems, then we must ask: What type of changes are required in the mutual perceptions of "Grand Strategies", for allowing co-operation? Which measures would have to be taken by either side if existing perceptions were to be changed intentionally in a direction representing greater trust? Dr. Garthoff has given us a very good basis for dealing with these questions. On both sides there is, of course, a multitude of views, a whole spectrum of perceptions and self-perceptions. But if we look at the extreme ends and at the center of the spectrum, we can very well identify the three schools of thought which he mentions: the essentialists, the mechanists and the interactionists. These schools exist on both sides. But their relative strengths may change. The relative strengths of these schools of thought on one side may depend very well on what the other side does because the young people who enter the scene are influenced by what they see. That is what matters. Dr. Garthoff has also described what the policy recommendations of these schools of thought are for preserving peace: The essentialists recommend confrontation, the mechanists recommend negotiations from a position of strength, and the interactionists recommend productive co-operation. Why is that so?

If one holds the perceptions of the essentialists then, obviously, confrontation is the best policy not only for one's own side but for the world as a whole. To explain this, Hitler is a good example. Had he met early confrontation then he might have acted differently, or he would have been removed by the German people. Co-operation encouraged him. Nevertheless, it seems obvious that for solving the global problems productive co-operation would be the best policy. If we had to choose confrontation, then it would be very difficult to cope with pollution of the environment, with hunger in the Third World, with terrorism, with the danger of war.

Co-operation could have many forms. René Herrmann yesterday sketched as a special form of co-operation a kind of Pax Russo-Americana in which both sides maintain stability by disciplining their alliances through confrontation. That is also a kind of co-operation. But I doubt whether it could be called productive. I think it would be very dangerous, leading, in the long run, to a division of the world into two camps. It would result in much fighting about the non-aligned countries of today. They would have to decide which of the two camps they should join. What the global problems need, obviously, is co-operation rather than confrontation. I think that is an objective diagnosis on which everybody can agree, be he essentialist or interactionist. Whether it is feasible to describe a viable road leading to co-operation is another question, and there the opinions will be divided.

The interactionists, of course, are already convinced that co-operation is possible. So the interactionists are not the problem, at least on the Western side. On the Soviet side, as our Soviet colleagues told us yesterday, also the interactionists among the followers of Gorbachev are not yet ready to grant to capitalism the capability to reform in such a way that it can be a partner indefinitely. They say they don't know yet, they have to study this question and reserve judgement for the future. But in principle the interactionists on both sides accept the possibility of productive co-operation. Grave difficulties for solving the world's problems would only arise if we had to assume that the arguments of the essentialists and the mechanists are correct, i.e. that in order to preserve peace the best strategy would be confrontation, or negotiation from a position of strength. The mechanists and the essentialists, of course, exist on both sides. So we should address their arguments and investigate to what extent these arguments are still valid, or whether some of them have been overtaken by the progress of history and are now obsolete. Moreover, we should look around whether there cannot be identified areas of overlap of perceptions and self-perceptions, and areas of overlap of self-perceptions and imagined self-perceptions, i.e. areas on which both sides agree and on which trust could be built. Common human values could be such an area, as indicated by Professor Dashichev and Dr. Buchholz. Another common basis could be the acknowledgement on both sides that also the other side is seriously concerned about the world problems of pollution, resource distribution, arms control etc. and is trying to make an honest contribution towards solving them.

We should discuss what it is that would have to be done in order to persuade the essentialists and the mechanists to give up goals which would have to be considered to be threatening to the other side. What could be done to persuade them to accept as feasible peaceful competition between the different social and economic systems? One thing that would have to be done is to learn the ideological language of the other side in order to avoid misunderstandings. In particular, the dialectic method is still widely used in Eastern Europe. For instance, as mentioned in chapter 15, Erhard Eppler, the chairman of the Committee on Basic Values of the Social Democratic Party of the Federal Republic of Germany, criticized Kurt Hager, the chief ideologist of the Communist Party of the German Democratic Republic, for continuing to use the term "enemy" for the capitalist system in a newspaper article and for denying the capability for reform and for peacefulness to that system. Kurt Hager did this right after representatives of the two parties had agreed that peaceful coexistence and even co-operation between the two systems is possible. This is just an example of dialectics, and I was surprised that Erhard Eppler was upset. It is indeed sensational that this kind of dialectic double-talk was given up by Gorbachev and his followers in the Soviet Union. From a marxist-leninist point of view there is nothing wrong with it. We shall have to learn that language, in order to understand the people who use it, and we must interpret it correctly, live with it, avoid risky conflicts, and cooperate whenever possible.

I think we should discuss what the essentialists and the pragmatic mechanists on each side would require the other side to do before they would be able to think that co-operation were possible. We should then see whether there aren't areas of overlap where there is a certain coincidence of perceptions and self-perceptions, areas that could perhaps be widened. Let us see whether we cannot agree that some perceptions are misperceptions. Let us see whether we cannot find compromises. We should not demand extreme steps like asking the Soviet Union to make her frontiers transparent and remove the Iron Curtain by tomorrrow, or asking the United States to introduce socialism and dissolve NATO. Setting such extreme conditions means that one does not want co-operation but believes in continuing confrontation.

On this small spaceship Earth we need co-operation. Therefore it seems to me that it would be desirable to find the perception correct that Gorbachev means business with his signals indicating that he accepts peaceful competition with the West and no longer insists on military confrontation. If we would send out signals indicating that we do not trust the change in Soviet security policy then, I am afraid, we would strengthen that group in the Soviet Union which also does not believe in the possibility of co-operation between the two systems. That would only be justified if we were firmly convinced that that analysis is correct and that in the long run there will be no alternative to a return to confrontation. If that were correct, we would have to prepare for it. But I don't think that assumption is justified. So far it seems that the risk of trusting Gorbachev is smaller than the risk involved in not trusting him.

Concluding remarks

Do we now understand the arguments, the experiences, the tracts of history which have led to the perceptions of the essentialists, the mechanists, and the interactionists, using the classification scheme introduced to this conference by Dr. Garthoff? Did we find any clue for distinguishing, in the real world, strategic and normative perceptions and self-perceptions from genuine ones, using the terminology applied by Michael Richter? Did we see areas of consensus between self-perceptions and imagined self-perceptions of the two sides? To what extent is there consensus about perceptions and imagined perceptions by the other side? Did we succeed, in some places, to identify misperceptions? What do we expect from each other in the long run? Which changes are required in our goals, in our perceptions of the goals of the other side, or in both, if long-term constructive co-operation between East and West is to become possible? Which obstacles can be identified, and what are the options for overcoming them?

We would have to go through the papers and discussions of this conference in search for answers. There is no doubt that a number of significant answers would be found, sometimes expressed in indirect ways. We would also be confronted with additional questions, and with answers which are still in a need of questions. Each of us may decide for himself or herself which kind of perception it was in each case that was presented. Probably, it was often a mix of genuine, normative, imagined and strategic perceptions and self-perceptions that was presented.

There is the suggestion by Professor Schulz and others that we should introduce some case studies. Topics mentioned were Iran, the Cuban Missile Crisis, Afghanistan, Vietnam. We might investigate the role that misperceptions and perceptions played in these cases. I quite agree with Professor Schulz that not every crisis can be explained by misperceptions alone. Probably no crisis can be explained just by misperceptions. But, as we have also seen during this discussion, hardly any crisis occurs without misperceptions. Misperceptions can be very dangerous. To try to put oneself into the shoes of the other side and understand what the motivations of the other side are is particularly important in times of crisis.

Another more general lesson of this conference is that there is no way to avoid all risks. Life in general is risky, and the East-West conflict is no exception to this rule. It is risky, as some participants pointed out, to fiddle around with stability in Europe by dismantling the East-West conflict. And it is risky to continue the conflict and the confrontation connected with it. Stability in Europe meant many wars and millions of victims in other parts of the world. But it is quite true that there is a certain risk involved in changing a situation which has prevailed for several decades. If new relations between the Eastern European countries evolve, and a new system is introduced in the Soviet Union, nobody knows what will happen. It is certainly risky. But, of course, it is also risky, particularly under these changes, to maintain confrontation. So whatever we do, there are risks. I think, however, that one can give a good number of arguments showing that the risk of supporting Gorbachev by assuming an interactionist attitude is smaller than the risk involved in not supporting Gorbachev in this way. We need a spirit of co-operation for tackling the many global tasks that we are facing: The gap between the rich and the poor countries, hunger, global pollution, the arms race etc. There should be more studies about these relative risks. Most people are influenced by their personal psychological background. They only see the risk of one policy. Some people consider it risky to change the situation, and some others say that it is very risky to continue. But nobody quantifies these risks. Studies of this kind should get more support.

With regard to what should be done in a tangible way Professor Senghaas gave a very good list of recommendations. My only reservation would be, as I already mentioned in the discussion, that these recommendations for changes in the military structure, economic co-operation, more contacts, a code of conduct, good as they are, will not convince the essentialists who believe in the value of confrontation. The interactionists, on the other hand, are already convinced and do not need such advice. The interactionists on both sides are not the problem, the essentialists are. They can, if at all, only be converted by an in-depth approach to the arguments supporting their adversary perceptions, not by mere recommendations for actions following from interactionism. So each side should study the arguments of the essentialists of the other side, and investigate the best way to deal with these arguments. It is not unlikely that the arguments for mistrust are not identical for Soviet and U.S. essentialists and that, therefore, the measures for taking care of them will have to be asymmetric for the two sides. This is another field for further studies.

17 Integrated Europe? [48]

A survey of European perceptions of Europe's role in regional and global political and economic co-operation [49]

If we want to move away from confrontation and improve the co-operation between East and West we must learn to understand and accept the long-range goals of our partners. If the perception of our goals by the other side seems to be incompatible with the security of that side, either our goals or the perception of our goals by the other side have to be changed. This, of course, is true for both sides. Last year, our research unit of the Max Planck Society organized a conference on this subject in which nine Soviet experts on Western affairs and ten U.S. experts on Soviet affairs, in the presence of 15 German scholars, told each other how they, and the people in their respective countries, saw the situation and the goals of their own countries as well as the situation and the goals of the opposite side. Of course, we had to distinguish the real, honest perceptions from strategic and normative perceptions which people just pretend to have, as well as self-perceptions and imagined perceptions. This exchange produced very interesting results. This year we had another meeting with the participation of representatives of the other countries of Eastern and Western Europe. [50] This meeting took place in March 1990 near Munich. It lasted three days and I should like to report here some

---

[48] This chapter contains a report on the perceptions, in March of 1990, of scientists and scholars from the countries of Eastern and Western Europe, of the United States and the Soviet Union (then still existing) of the political situation of Europe and of the risks and chances of future developments. Although the course of events of recent years has made many of these perceptions appear obsolete, we are reproducing this report because the meeting on which it is based is a good example of the type of "brainstorming" by scientific representatives of different backgrounds and nationalities that can call attention to potential dangers, to available options for coping with them, and to open questions that need careful analysis.

[49] Published in: J. Calliess (Editor), Der Neubau Europas. Loccumer Protokolle 19/90, Evangelische Akademie Loccum, 1991, pp. 754-762.

[50] K. Gottstein (Editor), Integrated Europe? Eastern and Western Perceptions of the Future. Campus Verlag/Westview Press, Frankfurt am Main, Boulder, Colorado, 1992.

of the findings which that meeting of 30 scholars from 13 countries produced.

First of all it was evident that considerable progress has been made in the change of perceptions of the goals of the other side. No longer is it accepted without question that it is the goal of the other side to "bury us", or to rule the world. There is now a clear erosion of the traditional image of an irreconcilable struggle between capitalism and socialism, in parallel with a spreading conviction of the interdependence of one world, the problems of which can only be mastered by joint efforts of mankind as a whole. Whereas only a few years ago spiritual coexistence with the West was considered to be impossible in the Soviet Union, the common belief in human values is now deemed to be the basis for co-operation.

At our conference at the beginning of June of 1989 it was still the East-West *conflict* which was foremost in our mind, and there was a lot of discussion on how this conflict could be controlled, overcome and turned into co-operation. After all, at that time, the GDR, Czechoslovakia, Bulgaria and Romania were still under regimes which are now usually called Stalinist. It still was appropriate, almost everybody agreed, to divide the spectrum of different views or perceptions of the other side into those of the essentialists, mechanists and interactionists. (The essentialists perceive the other side as an essentially hostile power which can only be dealt with by confrontation; the mechanists view it as governed by geopolitical considerations, reacting in opportunistic ways, so that negotiations should best be undertaken from a position of strength; the interactionists, finally, believe in change by rapprochement, in the existence of possibilities for peaceful competition and for constructive co-operation.)

In the meantime the events in Eastern Europe have certainly led to a further reduction in the number of essentialists on both sides. To the extent that they still exist they concentrate their mistrust, their misgivings and their hatred not so much on the government on the other side but on the conservative opposition forces which supposedly are waiting in the wings to take over when it becomes obvious that Gorbachev and his policies fail. In this sense the essentialists and their arguments still have to be taken seriously. Moreover, the old patterns of thinking still survive in another way. The increasing number of East-West contacts seems to lead to a strengthening of traditional perceptions. This happens because people in commerce and industry who have to take decisions in joint ventures or in making commercial offers do not know enough about the other side and are insecure about what to expect and how reliable or efficient the partner will be. In this situation they often use the old perceptions, and this leads in some cases to unauthorized intervention into the affairs of the other side, intermingled with deep skepticism about the chances of success for these interventions and that is for the chances of success of co-operation.

At the same time a new danger to stability has arisen: exaggerated "friend images". It derives from the over-optimism of some interactionists who expect co-operation to lead to immediate success in the economic field, but also in the fields of education, justice, international relations etc. Expectations that are unrealistic and cannot be fulfilled may lead to disillusionment, public frustration, disappointment and, as a consequence, to political instability. The perceptions underlying these unrealistic expectations may be called exaggerated friend images: The former opponents are seen as new friends who will help altruistically! This phenomenon has to be studied and taken into account with great care.

Generally speaking, it turned out that there is considerable agreement between the expectations which people in the countries of Eastern Europe have with respect to their new governments: Turning away from ideologies, introduction of efficient, lawful administrations free of corruption, transition from planned economies to market economies, integration into Europe. There are, however, significant differences about how to fulfill these expectations. Poland favours an uncompromising transition to market economy accepting social consequences such as an increasing gap between wages and prices and large-scale unemployment, hoping that the mechanisms of the market will improve the situation as time goes by. Other countries prefer a slow, controlled, step-by-step transition. It is interesting that it is not the socialists but conservative movements like the Democratic Forum of Hungary who favour limitations to a free market. U.S. participants express misgivings that such limitations might remove just those factors of a market economy which make that system work successfully, for instance by attracting foreign investment capital.

The Soviet Union is in a difficult situation because of the large number of dissatisfied citizens critical of the results of perestroika. The former power centers of the Central Committee, Council of Ministers and the Supreme Soviet with their huge bureaucracies still exist. But Gorbachev tries to establish a fourth supreme power center, that of the presidency, and these efforts do not meet with friendly sentiments from the old bureaucracy. This critical potential is enlarged by discharged military personnel, by party officials deploring their lost powers and fearing dismissal and by large sections of the population annoyed with rising prices, lack of consumer goods, crime, clashes between nationalities inside the Soviet Union, signs of dissolution of the Warsaw Pact, and the forthcoming unification of Germany on non-socialist terms. The latter fact, in particular, seems to indicate to the "man in the street" in the Soviet Union that his country which suffered so terribly in World War II was now losing the fruits of victory. It takes a lot of persuasion from supporters of Gorbachev like V. Dashichev to tell the people that Soviet soldiers in World War II did not die for the construction of Stalin's "geopolitical architecture" in Eastern Europe which led to the "Cold War", but for the liberation from fascism, and this goal was reached. To get rid of Stalin's legacy, Dashichev explains, would produce more gains than losses for the Soviet Union. But for a unified Germany to join NATO, if NATO remains a military alliance, would be quite unacceptable to the Soviet public.

In the discussions about the political structure of future Europe the German question plays an important role not only for the Soviet Union but also for the other countries of Eastern and Western Europe. In Poland, due to its painful historical experience, German unification is viewed particularly critically. So far Polish security rested officially on three pillars: friendship with the Soviet Union, the existence and the border guarantees of the German Democratic Republic, and the Warsaw Pact. What will be the policy of the Soviet Union towards Poland when the ideological element is eliminated? What will be the consequences for Poland when the Soviet Union and Germany establish cooperative relations? In Poland people hope that a new European security structure will be set up before Germany is unified. The Warsaw Pact should not be dissolved before NATO dissolves. Hungary, Romania and Bulgaria have a predominantly positive outlook on German unification. Particularly in Hungary there are no feelings of resentment towards Germany. Facing an enormous foreign debt, stability is sought by joint ventures and common projects of all kinds with the countries of the West. Special hopes center on a regional co-operation between Hungary, Austria, Yugoslavia and Italy. Unilateral dependence is to be avoided. Nationalism is a special danger for Hungary. Hungarians living abroad are the largest minority in Europe. Particularly the relations between Hungary and Romania are burdened by this historical legacy. Here is another confirmation of the old experience that regional conflicts are revitalized when global tensions are reduced. It is this phenomenon which is also feared in Poland.

There is no long-range planning in Romania. Foreign policy is reduced to pleas for foreign assistance. East and West are still seen as two separate worlds divided by a "prosperity curtain". For the work of reconstruction many Romanians would prefer France as their main partner, others plead for Germany. From a Romanian point of view the united Germany will take over from a retreating Soviet Union as "continental superpower" which should be controlled by NATO.

Scholars from the GDR believe that a longer transition period will be needed for replacing former antagonism between the systems by a new European order. It will be a difficult period because unification is proceeding faster than European unification. There is a certain risk that German unification will not be the end of the division of Europe. This could mean a continuation of confrontation. The German desire for integration into the West and association with the East could perhaps be brought closer to realisation if the troops of the four victor powers remained where they are. Perhaps East-West integration could start this way on German soil.

Recent German developments were not planned by the Soviet Union. They were a consequence of suspending the "Njets" of the Gromyko era. Now positive concepts are in demand, such as a new task for the Warsaw Pact. The Warsaw Pact could take on a new role as a consulting mechanism between governments - no longer parties - of Eastern Europe to prevent a "balkanization" of the region.

Whereas representatives of the Soviet Union, Poland and the GDR had sorrowful thoughts about the *political* consequences of German unification, it were the *economic* consequences which gave rise to skeptical contemplations of Western participants. An American scholar remarked that the Germans now might have found the formula how to structurally ruin the most stable economy and democracy in Europe. An Italian participant used the expression that also Germany will now have its Mezzogiorno. From the Netherlands came the request that not only the four victor powers and Poland but also the other European countries with economic interests in Germany should take part in the negotiations about German unification. The currency of the Netherlands, after all, is strongly coupled to the German Mark.

Great Britain with its traditional inclination to maintain a European balance of powers is concerned that the Soviet Union might crumble, that a unified Germany with its economic preponderance might get out of control, and that separate socio-economic developments might occur in Central Europe and within the present borders of the Soviet Union. Under these circumstances a strengthening, and at the same time a reform, of NATO is deemed necessary. The new institution emerging from this reform should give security to a unified Germany while keeping it within limits. The new institution should also be a negotiating partner for the Soviet Union which continues to be considered as a security risk.

France has hesitated for a long time to recognize the changes going on in the Soviet Union. Gorbachev was viewed as a Machiavelli. Today it is no longer his sincerity that is in doubt but his capability to succeed with his reforms. German unification is viewed with a certain uneasiness because French identity under the new circumstances has not yet been clarified. Will there be a renewed NATO with Germany firmly anchored in the West, or a collective security system from the Atlantic to the Urals, following de Gaulle's model, with a dissolution of the two blocs, formation of new alliances, minimum deterrence, and a centrally located Germany that has ceased to be a threat? In a new European Defense Community under the first model France hopes to play a leading role. But the increasing integration of France into a European identity with constraints for the status of France could cause frustration. The headlines of French newspapers during the visit of Mazowiecki and Jaruzelski to Paris gave the impression that the year 1938 had returned.

In Italy the developments in Eastern Europe and Germany are watched with a mixture of puzzlement and sympathy. From the Italian experience (South Tyrol) one believes that economic prosperity solves ethnic problems. German unification is supported but Germany should remain in NATO.

For the perception of Americans NATO is important. From the experiences of this century the average American believes that Europe cannot keep the peace without the presence of U.S. troops. That is why Europeans are willing to pay for these troops. But the troops would go home if Europeans - and particularly Germans - would want them to leave. On the other hand NATO grew in 35 years, and the developments in Eastern Europe happened only a few months ago. Grown structures cannot be changed overnight. Washington is now in a nostalgic mood. The world to which it was used is crumbling. Washington seemed to be the center of the world, now dramatic events are happening elsewhere, and Washington is only a spectator. But now that the negative task of containing communism is completed, the positive tasks can be tackled which so far had second priority. In co-operating with other countries the criteria of democracy can now be taken seriously, and policies can be designed in a differentiating way. A complete retreat of the United States from Europe is impossible because of the gigantic dimensions of U.S.-European economic relations.

In Switzerland neutrality is the Holy Cow. But can Switzerland afford to stay outside the European Economic Community? Détente devaluates the former role of the neutrals as honest brokers. Meanwhile, the Swiss military have no inclination to join agreements on conventional disarmament. Their argument: Disarmament should not be started by the neutrals. Their defensive capacity should be maintained until the others have disarmed.

This was just a selection of a few statements out of three days of discussion. Many questions remained open and are to be discussed at another conference next year. According to a classification by M. Richter they may be subdivided as follows:

1. Questions dealing with the ideological disputes in the Soviet Union.
2. Questions regarding the psychology, the strategies for change and the policies of the new elites in the countries of Eastern Europe, and their perception by Western countries.
3. How do the changes in Eastern Europe - and the fact that some things did not change - affect politics in the European context?
4. Is there a common normative basis developing between Western countries and the former Socialist countries?
5. To what extent do the hitherto existing perceptions of political and military threat shift to the economic sector? Which roads to a new economic integration, and which obstacles are perceived?
6. Will nationalism play a substantial role in future Europe, perhaps as a consequence of German unification or of minority problems in the South-East?
   Which perceptions could serve as a basis for potential agreement or for continuing differences?
7. Which safeguards could be built in interstate relations to promote the fading away of the remnants of ideological confrontation and to prevent their revival? Which obstacles for a new security order are noticed, and which options for overcoming them are perceived?

8. Will the removal of ideological differences lead to confrontations
on another basis (perhaps between the camps on both sides of the
"prosperity curtain")? Can this be prevented, e.g. by the
development of a common normative basis and the setting-up of
joint institutions?

In particular, the participants suggested:

- to identify the perceptions which brought about recent
changes in the Soviet Union, and to investigate whether
these perceptions are promoting or obstructing East-
West, East-East and West-West integration. Which
instabilities in East-East relations did they cause?
- to study the interactions between economic and political
relations,
- to do research on the perceptions of Germany's
neighbours regarding German unity,
- to look into the ideas on security with respect to growing
nationalism (New UNO Charter with improved
possibilities for sanctions?),
- to discuss, with participation of representatives of
commerce and industry, active East-West co-operation
for the stabilization of Eastern economies.

The great number of questions and suggestions shows the need and the
demand for international discussions of the type reported here. Many of these
questions are of great political significance but to answer them properly,
scientific investigations are required.

18 Perceptions of the Europe of Tomorrow [51]

Introductory remarks

We are talking about perceptions because there can be little doubt that the reality of politics, the potential for co-operation as well as the potential for conflict is not just determined by the so-called facts of life but by the perceptions of the facts of life. Decision-makers on all levels are not influenced by whatever the case may be, but what they *think* that the case might be, i.e. by their *perception* of events, ideologies, intentions, goals, etc.

We started this project during the period of Cold War before Gorbachev came to power. At that time, it was obvious that the arms race, the stalemate in arms control and disarmament negotiations, was due to mutual distrust, and the distrust, in turn, was due to the perceptions on both sides of the long-range political goals of the other side. These goals were perceived as extremely threatening. Both sides considered it necessary to deter the other side with nuclear weapons. The West concluded from its perceptions that the U.S.S.R. was planning to spread communism all over the globe by the use of force, and the U.S.S.R. assumed that the capitalist powers would destroy the home of socialism, i.e. the U.S.S.R, if they would get a chance to do this with impunity. Nobody was able to check whether the leaders on both sides really had these intentions, but the perceptions existed, and that was enough for starting and maintaining the arms race, and for preventing honest co-operation.

---

[51]This chapter contains an edited version of the introductory remarks by the author to a conference which was held in June 1991, a year and three months after the conference described in chapter 17. The full transcription of the proceedings of this new conference, including the discussions, in which competent participants from 17 countries (Eastern and Western Europe, the former Soviet Union, the United States and Canada) assessed the developments and visible trends, and perceived the actions by the political forces of their own countries and by foreign governments was published in: K.Gottstein (Ed.), *Tomorrow's Europe. The Views of Those Concerned.* Campus Verlag, Frankfurt am Main 1995 (845 pages). Moreover, the introduction by the author to the General Discussion of the conference is reproduced at the end of this chapter.

Stressing the significance of perceptions does not mean that facts and figures do not matter. But facts and figures are assessed, judged and weighed by perceptions, as is demonstrated by the observation that opposing parties often do not agree as to what the facts and figures are.

So it was not only worthwhile but necessary to study the observations and experiences on which these perceptions were based, and discuss the results with representatives of the other side, in order to find out whether it might not be possible by this process to remove some misunderstandings or "misperceptions", and develop methods of co-operation.

When we started this series of conferences a few years ago, it were the enemy images showing up in the perceptions that we were concerned about. They were an obstacle to peaceful co-operation for the solution of global problems, they were leading to confrontation. Then came the period of overoptimism after the announcement of perestroika and glasnost, and we had to warn against exaggerated friend images. Many people both in the East and in the West considered their former official enemies now to be friends, and harboured sometimes unrealistic expectations as to the fruits of the co-operation now possible. People in the East sometimes expect large-scale economic help which the West is only willing or capable to grant in a very limited way. People in the West are often hoping for the introduction of stable democracy and a functioning market economy in the East which it is impossible to achieve in the Soviet Union and in many East European countries in such a short time and without the necessary infrastructure and experience. So, all kinds of disappointment are in the offing which could lead to renewed animosities. To this situation are added the old rivalries between individual East European and Central European nations and populations, between Hungarians and Romanians, between Armenians and Aseiris, between Serbs and Croats, to name only a few. At a time when Western Europe is in the process of giving up voluntarily national sovereignty in favour of a central administration in Brussels, the Eastern European countries, after a long period of involuntary membership in a centrally administered bloc, are longing for independence. But in many cases, this means revival of ardent nationalism. So, on a different level, we observe perceptions again that lead to enemy images. We shall have to study their origins and must look for ways and means for overcoming them.

There is another dangerous risk that has to be kept in mind, and studied. Foreign Minister Genscher called attention to it recently, warning against it. It is the risk that events in the Soviet Union could prevent the U.S.S.R. from joining the common European House so that a new wall would be erected along the Eastern border of Poland. Some say that the introduction of democracy and market economy would be easier when the Soviet Union has a strong central government. Some claim this should be left to the individual republics.

Economic co-operation between East and West in Europe is also made difficult by misunderstandings about the term "market economy". In the Soviet Union, in parts of Eastern Europe and also in parts of the former GDR, market economy is often understood as unfettered liberalism of the Manchester type. But, of course, Western countries have long ago given up this type of liberalism for one limited by social legislation, and their social and economic stability is due to this limitation. This social net is still under construction in Eastern Europe, and that will take time. This is also true for the removal of the still existing "prosperity curtain".

If we want to learn from each other, we have to ask questions. Michael Richter compiled a list of questions the answers to which can be perceived as being critical for the determination of the future of Europe. Let me just quote some of them in a slightly modified way:[52]

- What will happen to the Soviet Union? I might add here: What will happen to the European Community? Who will be allowed to join, and when?
- What is the psychology of the new leadership in Eastern European countries, and how is it received in the West?
- Will there be a common normative basis in East and West? What are the values that are cherished most on both sides?
- Are there new threat perceptions in the economic area? By what kind of co-operation could they be possibly overcome?
- How is nationalism to be dealt with?
- Which options are there for a new security order for Europe?

I think it will be very interesting to discuss the perceptions that can be found in the Soviet Union, Eastern Europe, Western Europe, the United States and Canada with respect to the significance of these and other questions and with regard to the various answers that are offered, in political debates as well as in scholarly literature.

---

[52] Compare the original list in chapter 17.

Introduction to the general discussion

In this Conference many different aspects of the "Europe of Tomorrow" were discussed. In addition to what we heard about the views on security, on economic co-operation, on regionalization, on Soviet reforms, on conversion, on nationalism, human rights and about the cultural and philosophical aspects, we must remember what was said regarding perceptions of East-West relations, the roles of the United States and the Soviet Union in Europe, the views on how to organize European integration and the perceptions in individual countries or groups of countries of their specific roles in Europe.

Is it possible to draw some general conclusions from this multiplicity of facts and perceptions regarding the future of Europe? If one were forced to put the problem of Europe in a nutshell then one might have to say that the problem which Europe now faces is how to harmonize the integration of Western Europe with the disintegration of the Soviet Union, and of the Warsaw Pact in general. Will it be possible to build some sort of common European structure in which the Soviet Union - either through a central government in Moscow or through its European republics - has some role to play? A common structure which allows to find some kind of common European stance with respect to the turmoil in the Arab world and the Third World in general?

We have heard the view of Professor Kaltefleiter that this is not possible. When he talked about the role of Europe in the international system, he had in mind a prosperous, democratic Europe which - together with the United States - could act as a haven of stability and project its benign influence under the umbrella of some kind of Pax Democratica into the less fortunate parts of the world. And the Soviet Union, in this perception, was considered to be one of those less fortunate parts. Indeed, its problems were seen as so immense that any assistance by Europe, the United States and Japan combined would be of no significance. This is certainly only *one* German view and not the view of *all* Germans. But also Professor Kremenyuk said that the reform of the Soviet Union will take a lifetime. We cannot expect quick results in democratization and economic recovery. But what is the consequence? Can Western Europe remain an island of stability and prosperity when there is upheaval on its Eastern border? Can Western Europe remain unchanged when the world around it changes? I do not think this is possible. We cannot put barbed wire on the prosperity curtain. The unification of Europe, and later of the globe, is a process that cannot be stopped because it is the result of modern technology, of modern means of communication and transportation. When there was the Iron

Curtain, it was possible for Western Europeans to ignore to some extent what happened behind it. But now the borders are open.

We did not discuss very much how Germans perceive the unification of Germany. If we had, we would have learned what unification means: that communists are now members of the German parliament, that troops of the Bundeswehr and of the Soviet Army are stationed on the same territory, that institutions and production plants designed under socialist auspices now have to be absorbed, under enormous difficulties, into a structure that was built under Western conditions. Germany is changing. It is not the old FRG any more. History offers many examples of similar changes. Italy changed tremendously when it was united in the last century and when its northern parts and its southern parts were joined which had been under very different rulers and influences for centuries - Austrian, French, Spanish, Arab - which had developed very different traditions, so that even now Northern Italy and Southern Italy are sometimes seen by Italians to be two different societies.

So, no doubt, Europe will change when it unites, just as the united Germany is undergoing change even though the government tries to continue its traditional policies. Let us prepare these changes on the European scene in such a way that catastrophes are avoided. And with special reference to the relations between Europe and the Soviet Union, I remind you of what Professor Kremenyuk said: "Let us develop a programme for the next 10 to 15 years which is feasible and which is acceptable to the Soviet Union as well as to the other European nations." We cannot ignore each other until the Soviet Union has solved all its internal problems. This could be very dangerous because the Soviet Union, due to its size, has an enormous potential for destabilizing Europe whether it is inside or outside its political structure. An unhappy elephant can severely damage a fragile building just as well from the inside as from the outside.

In which way could a programme of co-operation between the Soviet Union and the rest of Europe for the next 10 to 15 years, as mentioned by Professor Kremenyuk, be approached? It should be a programme of solidarity and of generosity as well as one of prudence and of clear-sightedness, a programme that does not create enemy images but perceptions of solidarity and of mutual assistance between equal partners. I hope that this conference will contribute a little bit towards strengthening such international solidarity in view of the enormous problems, not only European but global, that are facing us.

19 International "Security in a Wider Sense" [53]

Certainly the topics of military security will continue to require the highest attention. Nuclear weapons are still around in huge numbers, and under less strict control, in the former Soviet Union, than during the Cold War. Warfare with many victims between Armenia and Azerbaijan, civil war in Georgia, disputes between Russia and the Ukraine regarding possession of the Crimea and the Black Sea Fleet give rise to deep concern, as already mentioned in chapter 2. Historical experience shows that nations as well as politicians are ready to resort to arms when they believe that their interests are at stake and when one of the opponents feels that armed fighting will give him a chance to win or to avert defeat. Warfare between the republics of the former Yugoslavia is another example.

Supporting the scientific community in the former Soviet Union is already a contribution to international security in a wider sense. Professor Salvini mentions additional topics related to international security in such a wider sense and requiring urgent attention: the "Ecological Bomb", the "Demographic Bomb", and the psychological foundations of the overriding principle of altruism.[54] I agree that, quite apart from the ethical aspect, altruism is a wise policy even from a selfish point of view. Psychologists in our academies may expound this in scientific terms. One might also look at the problem by realizing that in our age of modern communication and transportation the poor and the rich on this globe now all belong to the same community. By supporting the poor we are supporting members of our own community and thereby contribute towards the health of the system in which we live. This is a scientific task and is not in contradiction to the rule that scientists should clearly distinguish between their statements, as experts, on what the facts are, and their statements, as politically interested citizens, on what should be done. Scientists, in their capacity as experts, are able to pronounce what should be done only if a

---

[53] Edited version of a contribution to the 5th International Amaldi Conference, "International Security in a Transformed World", Heidelberg, July 1-3, 1992.
[54] Giorgio Salvini, The Future of the Amaldi Conferences. In: Proceedings of the 5th International Amaldi Conference of Academies of Sciences and National Scientific Societies, "International Security in a Transformed World", Heidelberg, July 1-3, 1992, pages 333-334.

clearly defined goal is given, like putting a man on the moon. They are also able to say, what should *not* be done if certain consequences are to be avoided, like heating up the earth´s atmosphere, or creating hatred between peoples.

A few days ago, at the celebration of his 80th birthday, Professor Carl Friedrich von Weizsäcker said "There is no problem that could not be solved by reason applied co-operatively". However, what are the obstacles to applying reason in a co-operative manner? This, too, can be investigated by applying reason cooperatively. But will reason lead to conclusive, unambiguous results? We have shown in chapter 2 that independent thinking can lead to highly differing results. In general, the multiplicity of views, interests and traditions results in strife when groups representing them get in touch with each other. Strife and fighting can be avoided, however, if the opponents have come to belong to a social system with a strong central authority that sets rules for a peaceful settlement of conflicts, and establishes enforceable sanctions against violations of the rules. It will be a task of paramount importance to design a new system with acceptable and enforceable rules for the peaceful settlement of the unavoidable conflicts that the future holds in store. The nations of both Eastern and Western Europe, and as a matter of fact those of all other parts of the world, require a new sense of partnership. They all face the huge challenges of our time which were described in the earlier chapters of this book and which can only be mastered by international and interdisciplinary co-operation. This will require joint institutions, not only for mediation, jurisdiction and conflict settlement, but also for joint international and interdisciplinary investigation of the problems involved. As mentioned in chapter 2, it will be necessary to form international interdisciplinary committees of scientists, engineers and scholars (including, if necessary, economists, psychologists, historians, political scientists, international lawyers) who devote themselves in an impartial way to the task of developing options for political action in the cases under consideration. The academies of sciences and the corresponding national scientific societies dispose within their memberships of the knowledge, the multidisciplinarity and the international connections that would be required for setting up the interdisciplinary and international framework needed for the purposes described. They should not shrink, therefore, from this task.

How can this requirement be operationalized? I suggest that, in parallel to the preparations for the next Amaldi Conference on problems of arms control and military security in the narrower sense - including questions of nuclear weapons, chemical and biological weapons, arms trade and non-proliferation, verification and conversion -, an international fact-finding commission be set up to study possible contributions of academies and scientific societies to the solution of problems of security in a wider sense. East European scientists, including scientists from the republics of the former Soviet Union, should be involved here. They could make valuable contributions and, in the sense mentioned above, their participation would strengthen their position at home. As we all know, the security of humankind is not only threatened by weapons of mass destruction but also by the "bombs" mentioned by Professor Salvini in the fields of ecology and demography. As mentioned previously, the recent United Nations Conference on Environment and Development (UNCED) at Rio has just supplied a summary of some of the open questions. The commission to be set up by the academies and scientific societies could state, on the basis of the best knowledge available, where we stand and what the risks are, under different assumptions about developments and about the action taken. Different options for action could be described, with the probable consequences.

This in itself is, of course, a very ambitious programme which could be the subject of a separate meeting of academies. Whether such a separate meeting should be called Amaldi Conference or given another name is of secondary importance. On the other hand, the name "Amaldi Conferences" has by now become known for meetings of academies on scientific questions of political relevance. It has finally become accepted that academies and scientific societies are able to contribute, because of their technical expertise, to the discussion of security questions, and that this is respectable and should be supported. Certainly Edoardo Amaldi would have supported the idea of looking at international security in a wider context. Of course, taking up new subjects of security in a wider sense should not detract from the very important subject of military security "in the narrower sense", as it was discussed here and in earlier Amaldi Conferences. New subjects should be taken up by new experts and discussed separately, and by no means at the expense of arms control questions. Perhaps, in the long run, there will be several permanent committees set up by the international community of academies and national scientific societies, just as the U.S. National Academy of Sciences has the Committee on Science, Engineering, and Public Policy (COSEPUP) in addition to its Committee on International Security and Arms Control (CISAC), as described in chapter 1.

20 Science Advice to the United Nations [55]

Introduction

Global coordination is needed if environmental and other catastrophes of a global character are to be avoided in the years to come. This means that the United Nations must be strengthened, and that interdisciplinary scientific advice of top quality must be made available to decision-makers at the global level. In this context it is of interest to take stock of the existing advisory mechanisms of the United Nations. Although they are mostly not coordinated, usually serving one single agency and its missions, they play a useful role and could serve as contributors to, or "building blocks" of, an overall interdisciplinary, electronically connected, advisory system for the avoidance of future catastrophes.

The remaining part of this chapter was written as a contribution to the Tenth International Amaldi Conference, held in Paris, 20-22 November 1997, at the invitation of the Académie des Sciences.

1 Why decision-making at the global level, based on scientific advice, is now required

When the United Nations (UN) were founded towards the end of World War II, threats to peace were only perceived in the military sense, and by individual nations against other nations. To cope with such threats, the Security Council was set up. The information it needed for its deliberations and decisions was only of a political and military nature. This changed soon after the end of World War II when the membership of the UN was widened to include many countries from the so-called Third World. Bridging the gap between the standards of living in these "developing countries" and the established industrialized countries was now considered to be of high priority for the preservation of peace, justice and equity. At the same time the two

---

[55] Published in: The Path to a More Secure World. Proceedings of the X International Amaldi Conference. Institut de France, Académie des Sciences, Paris 1998.

opposing blocs in the Cold War were competing with each other in trying to obtain allies in the Third World by aiding the new member states in their efforts to industrialize. Science and technology were seen as instruments of development. All this required information of an economic, social, technical and scientific nature, in addition to the traditional political and military intelligence. The efforts to obtain this type of information at the level of the UN will be described in section 2 of this chapter. Of course, it was not only at the global level of the UN that such information was needed for sound decision-making. After all, the introduction of the Marshall Plan, the creation of the Bretton Woods institutions, the first beginnings of a European Community, and the conclusion of new social agreements in the industrialized countries also characterized the first decade after the great war and raised problems at regional and national levels which forced governments to seek scientific and scholarly advice.

It is interesting to compare the situation in the late 1940s and early 1950s with that of half a century later. *The Report on Human Development of the United Nations Development Programme (UNDP),* published after the end of the Cold War in 1994, states that "security" is no longer seen under the aspect of concern about aggression against national borders. When security of people is threatened anywhere in the world, all states may be affected. Famine, ethnic conflicts, decay of social structures, terrorism, environmental pollution and drug traffic are no longer isolated events within national borders. Their impact is felt worldwide due to modern means of communication and transportation. It is less costly and more humane to tackle the causes of these threats than to treat their consequences. For most people feelings of insecurity result from concern about job security, income security, health security, environmental security, security against crime, and other aspects of daily life. A new concept of human security is therefore required, the Report says. It should have the individual at its centre, and protect the life chances of future generations as well as those of the present ones. This also means that nature, on which all life depends, is protected. A new era of co-operation for development among all states of the world is to be initiated, the Report further suggests. The goal must be economic partnership and co-operation. The affluent states should be ready to pay the poor ones for services rendered in the global interest, such as environmental controls, the control of drug production, the fight against infectious diseases, the destruction of nuclear weapons.

Some of today's global problems, which are worrying to an increasing degree both the public and the politicians in the industrialized countries as well as in the so-called Third World were listed in the "window session" of the 9th Amaldi Conference held in Geneva under the auspices of the United Nations and the European Organisation for Nuclear Research (CERN), 21 - 23 November 1996:[56]

- pollution of soil, water and air
- destruction of the ozone layer
- heating of the atmosphere
- desertification
- disappearance of animal and plant species in alarming numbers
- the human population explosion
- food and energy shortages
- migration of millions of people
- nationalism, racism and ethnic "cleansing"
- psychological, social, and economic instabilities
- civil wars and weapons trade
- the threat of nuclear proliferation and of the misuse of nuclear materials,

If humanity is to come to grips with these problems, measures will have to be decided upon at the global level, and they will have to be implemented and controlled at the global level. There is a growing realization that humanity's needs cannot be met entirely at the national level. This and the spontaneous *de facto* globalization in economics, trade and communications cause a trend toward "world governance". There are divergences of views as to what "world governance" should mean: *"for some, world governance refers exclusively to the United Nations and its activities, while for others, regional organizations and the whole array of international organizations and linkages are included."*[57]

---

[56] K.Gottstein, The role of national academies and scientific societies in supplying advice on the nature of global problems and on the available options for coping with them. Introductory Remarks. In: Proceedings of the IXth International Amaldi Conference of National Academies of Sciences and National Scientific Societies, Geneva, 21-23 November 1996, Accademia Nazionale dei Lincei, Rome 1997.

[57] Report on Working Group 5, 47th Pugwash Conference on Science and World Affairs, Lillehammer, Norway, 1-8 August 1997.

It seems to be obvious that the solution of problems that have to be tackled at the global level is a task for the United Nations. But is the UN, with its present means and structure, able to handle this task efficiently? In the Reform Commission for the UN the idea is being discussed to set up, in analogy to the existing Security Council, a Council for Environmental and Economic Security with strong executive power, and to incorporate the World Bank and the International Monetary Fund into the UN so that their advisory apparatus becomes part of the UN advisory system.[58]   The foundation of a World Environment Council to meet ecological challenges was also proposed by Chancellor Helmut Kohl who stressed the connections between ecology and economy and the obligation not to destroy the "treasure of nature".[59] It is striking that the UN has a High Commissioner for Refugees but no High Commissioner for the Protection of Environment, Climate, Biodiversity and other equally important tasks.

Before we address ourselves   to the question of how such an interdisciplinary and international advisory body for the UN could be organized, let us have a brief look at some previous efforts to supply the UN with scientific advice.

---

[58] Dr. Richard von Weizsäcker, private communication, June 1997.
[59] Süddeutsche Zeitung, 12./13. August 1995.

2   Scientific advice to the United Nations in the past. Some examples

2.1 *ACAST, CSTD, and the UN Conference on Science and Technology for Development*[60]

In 1961 the *Economic and Social Council (ECOSOC)* of the UN decided to organize a *Conference on the Application of Science and Technology for the Benefit of the Less Developed Areas*. This conference took place in Geneva in February of 1963. It had the character of a large fair for science and technology. 1665 persons participated, almost two thousand papers were submitted describing a great number of scientific-technological projects of potential interest to developing countries. There were no decisions on programmes or practical steps, and so the conference of 1963 remained without significant consequences for the countries of the "Third World".

Nevertheless, attention had been called to the development factor "science and technology", and this led to some institutional changes within the apparatus of the UN:

- In August of 1963 ECOSOC decided to create an *Advisory Committee on the Application of Science and Technology to Development (ACAST)*. The members of ACAST, about 30, were appointed in their personal capacities after consultation with the governments. ACAST was given two tasks:
    a. to identify problems deserving special research efforts
    b. to produce a "World Plan of Action"

---

[60] English translation (by the author) of parts of: K.Gottstein, Wissenschaft und Technologie für die Dritte Welt. Die Konferenz der Vereinten Nationen über Wissenschaft und Technologie im Dienste der Entwicklung - Wien, 20.-31.August 1979 - und ihre Rahmenveranstaltungen, Max-Planck-Institut zur Erforschung der Lebensbedingungen der wissenschaftlich-technischen Welt, Starnberg 1979 (unpublished). See also:
- Subas C. Pati, Experiences with the Vienna Programme of Action, Forschungsstelle Gottstein in der Max-Planck-Gesellschaft, München 1986
- K. Gottstein, Science and Technology for the Third World: The United Nations Conference on Science and Technology for Development, ECONOMICS, Vol.21, 1980, p.134 ff.

- In 1964 an Office for Science and Technology (OST) was set up within the UN Secretariat. One of its tasks was to serve as a secretariat for ACAST.
- The Administrative Committee on Coordination (ACC) which mediates between the UN and its sub-organizations and special organizations gave itself, after the conference of 1963, a sub-committee for science and technology which existed till 1978.

ACAST fulfilled its mission. However, its list of problems and its "World Plan of Action" failed to attract the expected attention. There was no link to authorities in charge of carrying out plans of action. In order to make progress in this respect, ECOSOC (consisting of representatives of 54 governments) in 1971 created an *Intergovernmental Committee on Science and Technology for Development (CSTD)*. This Committee, following a suggestion by the Secretary-General of the UN (drafted by a group of experts), suggested to plan a new UN Conference on Science and Technology for Development which, in contrast to the conference of 1963, should concentrate on concrete problems of development, including political, financial and organizational problems. For the determination of the organizational structure and the programme of the conference an *Intergovernmental Working Group (IWG)* under the chairmanship of a Brazilian diplomat was set up. IWG recommended that the conference should deal with the following topics:

1. science and technology for development,
2. institutional arrangements and new forms of international co-operation in the application of science and technology,
3. utilization of the existing United Nations system and other international organizations,
4. science and technology and the future.

The UN General Assembly accepted this proposal. The CSTD was converted into a Preparatory Committee, and Vienna was chosen as the venue of the conference. Preparations were to be undertaken at national, regional, and global levels. For the illustration of the application of science and technology to development the following areas were chosen by the Preparatory Committee:

- food and agriculture
- material resources and energy
- health, human settlements, and the environment
- transport and communications
- industrialization

130 countries prepared national papers on these subjects and submitted them to the conference which, after a number of regional meetings in Arusha (Tansania), Mexico City, Geneva, Bangkok, Beirut, Bucharest, Panama City, Cairo and Amman, took place in Vienna, 20-31 August 1979.

It was a conference of governments. They were mainly interested in the political, organizational and financial aspects of the application of science and technology to development. The members of ACAST, and scientific circles in general, had soon realized that it would hardly be possible in this setting to discuss questions of scientific and technological substance. For this reason ACAST organized a separate *"International Colloquium on Science, Technology and Development"* which was also held in Vienna, 13-17 August 1979, i.e. right before the official conference. In order to prepare the agenda of the ACAST Colloquium, preparatory conferences had taken place, with the assistance of OST, in Morocco, Estonia (then part of the USSR), Jamaica, Singapore, Malaysia, Mexico and Germany (West-Berlin).

371 scientists and representatives of scientific organisations from 95 countries took part in the colloquium. They met in 14 Working Groups and in plenary sessions. As an introduction to the discussions an initial report had been prepared for each working group. At the end, each group presented a report to the final plenary session, containing recommendations and general remarks about the subject area discussed. Apart from special topics, such as food and agriculture, information systems in science and technology, etc., there was, already at this 1979 colloquium, a group dealing with "interrelations between science and technology and global problems". One of the statements issued may be of particular interest in the present context:

Scientists can assume five principal functions with regard to the application of science and technology to the development process:
- *draw international attention to global problems*
- *store, retrieve and exchange data*
- *generate new knowledge by means of research*
- *coordinate evaluation systems*

- *re-orientate development strategies, design alternative patterns of development and pursue exercises in global modelling*

The ACAST Colloquium directed a Message to the UN Conference which contained a great number of recommendations for measures to be taken in the fields of agriculture, health services, natural resources, transport, communications, industrialization, appropriate technologies, human settlements, environment, energy, population policy, and information systems, in addition to some general comments. The Conference itself agreed on the "Vienna Programme of Action" which covers 32 printed pages and is divided into 100 sections with many sub-sections. Its recommendations are mostly of a political and organizational nature and concern the substance, mechanisms and structures of a scientific and technological policy for developing countries, questions relating to the transfer of technologies, international co-operation, information systems, training and financing, and the role of the UN in science and technology. Suggested are the establishment of an Inter-Governmental Committee on Science and Technology for Development (IGC) open to all member states of the UN, the preparation of a long-term system for financing scientific and technological activities in developing countries, and the establishment of an international global information network for the special needs of developing countries in science and technology, providing also data on socio-economic, legal and other aspects required for decision-making. IGC was established as the direct subsidiary organ of the General Assembly. Moreover, a new Advisory Committee on Science and Technology for Development (ACSTD) was constituted composed of 28 eminent personalities from all the regions of the world, in order to ensure adequate and effective provision of scientific and technical advice for the deliberations of the IGC. Furthermore, a Center for Science and Technology for Development (CSTD) was established as a new and distinct entity of the UN Secretariat. The Centre's main responsibilities are the provision of substantive secretariat services to the IGC and the ACSTD. CSTD was also assigned the task of interacting with governments, particularly through their designated national focal points and with inter-governmental and non-governmental organizations active in the field of science and technology for development.[61]

---

[61] The United Nations Centre for Science and Technology for Development. Responsibilities, Tasks and Organization, New York 1985.

Many of the suggestions made by the UN Conference, and particularly by the ACAST Colloquium, were based on very good reasons and are certainly worth being considered. Why, nevertheless, most of them were not implemented, will be discussed in section 3 of this chapter. In the meantime, ACSTD and CSTD have ceased to exist.

2.2 The United Nations Conference on Environment and Development (UNCED)

This Conference took place in Rio de Janeiro in 1992. In resolution 44/228 of 22 December 1989, the General Assembly of the UN had decided that UNCED, in addressing environmental issues in the developmental context, should have among its objectives "to promote environmental education, especially of the younger generation, as well as other measures to increase awareness of the environment".[62] Obviously, environmental education requires the availability of scientifically reliable information that can be taught. Several organisations felt responsible for the preparation of such material. The World Health Organisation (WHO) of the UN, for instance, in early 1990 established a Commission on Health and Environment for this purpose. It is interesting to look at a statement which the representative of the Netherlands in the Preparatory Committee for UNCED made on behalf of the European Community regarding the work of this WHO Commission:[63] "This Commission has made an inventory of our knowledge of the impacts of environmental changes on human health, indicated areas where further research is needed, and laid the basis for WHO to develop strategies to tackle the problems of health and the environment in the future. ... the final conclusions and recommendations of the WHO Commission should be

---

[62] U.N.General Assembly, Preparatory Committee for the United Nations Conference on Environment and Development, Document A/CONF.151/PC/21 of 28 January 1991.

[63] Third Session of the Preparatory Committee for the United Nations Conference on Environment and Development (Geneva,12 August-4 September 1991). Statement by the Representative of the Netherlands on behalf of the European Community and its Member States. Plenary. ENVIRONMENTAL HEALTH ASPECTS OF DEVELOPMENT/ EDUCATION AND PUBLIC AWARENESS/POVERTY AND ENVIRON-MENTAL DEGRADATION. Presidency of the European Communities, Nederland 1991.

considered for inclusion in Agenda 21 ... Four main areas have been identified for consideration for Agenda 21. They are:

(a) Basic Health Needs Related to the Environment;
(b) Control of Communicable Diseases;
(c) Meeting the Urban Health Challenge;
(d) Reducing Health Risks from Environmental Pollution and Hazards.

... we think that the element of coordination deserves our attention. For each of the main areas ... just mentioned, it requires **intersectoral co-operation and coordination, which in many countries today are lacking, fragmented, inadequate or ineffective**. National actionplans must therefore be developed. Co-operation between on the one hand the various ministries involved, and on the other hand the governmental organisations and non-governmental organisations is essential in this respect. ... **We believe that sustainable development requires an interdisciplinary, holistic approach, where economic, ecological and social aspects are being considered in an integrated manner at the local, national, regional and global level. Therefore sustainable development must be translated into concrete national policies, that simultaneously address ecological sustainability, equity, the introduction of population policies and an increase in productivity and economic growth in order to alleviate poverty, in particular in developing countries. Implementation of these policies requires effective institution building and training**". (Bold type by K.G.)

"Institution building" by the international scientific community is certainly required for the preparation and delivery of much-needed scientific advice to governments with respect to the task of dealing with the well-known threats to security in the wider sense, including environmental security. An "Urgent Appeal from the *Association of European Universities (CRE)*" contains some interesting statements in this connection:[64] "...We are presently living in an interactive, dynamic and often turbulent world. Systems are multifaceted and

---

[64] Urgent Appeal from the Association of European Universities (CRE). Preparatory Committee for the United Nations Conference on Environment and Development (Third Session: Geneva, 12 August - 4 September 1991). Standing Conference of rectors, presidents and vice-chancellors of the European Universities, Geneva, August 1991.

complex. We must act locally and globally, often in conditions of great uncertainty. Frontline and fundamental knowledge is required in technology, economics, biology, physics, chemistry, law, medicine, management - and we must be able to apply this knowledge to all human activities and global processes. These include energy production and consumption, greenhouse gases and climate, transportation, communication, trade, agriculture and soil erosion, water management, waste disposal and the recycling of materials, life style changes, changing relationships between poor and rich nations, etc. **Universities (by "universities", we understand all institutions of higher learning and research) are the only institutions which have the comprehensive knowledge needed today. They must therefore work with governments and business (by "business", we understand trade, commerce, industry and services) to apply this knowledge to high priority projects and programmes. Developing this research capacity will require a new dimension of cooperative efforts within countries, regions and on a global scale, and universities will work hard to promote such co-operation. ..."** (Bold type by K.G.)

There is no space here for summing up the results of UNCED in Rio which dealt with "the most complex and difficult problems mankind has to face: local, regional and global deterioration of the environment (air, water, soil); stability of global climate vis-a-vis the greenhouse effect; demographic explosion, especially in the poorest countries; the dissipative use of finite natural resources; the disappearance of animal and vegetal species menacing the biological richness of earth; deforestation, desertification and degradation of agricultural land; the large, and often increasing gap between rich and poor countries; the unequal distribution of scientific and technological knowledge, a wealth maybe more important than material wealth".[65] UNCED produced valuable documents. The Agenda 21, with its 40 chapters and 115 special topics on roughly 800 pages is a useful collection of detailed information for many questions of environmental and developmental policy with plans of action and first estimates of the finances required for their implementation.[66] The Conference, with its "Rio Declaration" of 27 principles and its Agenda 21, raised the awareness of the world public for the environmental problems of the planet Earth. Moreover, the Conference set up an international body, the

---

[65] Umberto Colombo, 5th Amaldi Conference, Heidelberg, July 1992.
[66] See: Barbara Unmüssig, Zwischen Hoffnung und Enttäuschung. Die Konferenz der Vereinten Nationen über Umwelt und Entwicklung: eine erste Bewertung. Vereinte Nationen 4/1992, p.117 ff.

Commission on Sustainable Development (CSD), to monitor the implementation of UNCED's decisions. Fifty-three member-states of the UN are represented on CSD.

What was achieved during the five years after Rio? Two conventions were launched, one on climatic change, and one on biodiversity, supplemented later by a convention to combat desertification. That may not be enough, but it is better than nothing. Nevertheless, it must be said that the global state of environment and development did not improve during this half-decade. When it changed, then - on the average - it deteriorated. The Summit Meeting of the UN in New York in June 1997, which was to be a follow-up to UNCED, ended without solid results. While fully recognizing, in drastic words, the consequences of global heating, President Clinton did not give firm pledges for the reduction of greenhouse gases in the United States, which contributes about one quarter of the greenhouse gases produced worldwide. There is opposition to stricter environmental standards by U.S. politicians mainly concerned with employment and the economy who fear an economic depression as a consequence of protective measures for the environment.[67] This is another indication for the necessity of an all-encompassing, multidisciplinary approach to the global problems, an approach which takes into account economic, psychological and other aspects as well as the entirely scientific arguments which many people do not understand (see also chapter 3). Encouraging, however, is the wide publicity which the global problems, and the different proposals for their solution, now receive, and the beginning preparedness of scientific bodies to devote their knowledge and their resources in a systematic way to the task of advising governments and the United Nations in this respect. The Appeal from the Association of European Universities, mentioned above, is an example.

2.3   The United Nations University (UNU)

A source of scientific and scholarly advice to the United Nations, as well as to the world community, is the United Nations University (UNU). "The Charter of the UNU was adopted by the General Assembly in December 1973. ... The UNU began operating in September 1975. The University works on the pressing global problems of peace and governance, environment, science and technology, and development. ... With a broad mandate and using a

---

[67] Süddeutsche Zeitung, 28/29 June 1997.

multidisciplinary approach, the University strives to ensure broad representation of the experience and viewpoints of various regions, cultural traditions and schools of thought. The subject of UNU's work ranges from development economics research to the application of science and technology in developing countries. ...

... Serving as the link between the world's academic community and the UN system, the UNU promotes multilateral activities, enhances the collaborative efforts of scholars in understanding the nature of global changes and broadens access to knowledge and information, particularly in developing countries.

The UNU is unique in its approach to the advancement of knowledge. It has no students in the usual sense, no faculty and no central campus. Its work is carried out through research and postgraduate training networks in both developed and developing countries. It functions as an autonomous organ of the UN General Assembly and is not an intergovernmental organization as typically conceived. The academic freedom guaranteed by its Charter enables UNU scholars and scientists to collaborate freely.

Although the UNU's Headquarters is in Tokyo, its growing system of research, training and dissemination work ensures that its activities are worldwide in scope.

... The University Council, made up of 24 members who are appointed jointly by the UN Secretary-General and the Director-General of UNESCO for six-year terms, sets the University's principles and policies. ...

The University is organized on a networking principle with the University Headquarters in Tokyo acting as the hub and coordination point of the worldwide system. Research and training centres and programmes are set up to work on specific problems requiring a sustained effort. ...

Financial support for the UNU comes entirely from voluntary contributions from governments, bilateral and multilateral agencies, foundations and other public and private sources. It does not receive any funds from the regular budget of the United Nations. Major contributions are made to its endowment fund, a capital fund invested to yield the basic annual income. The UNU also receives annual operating contributions and specific programme and project support. As of December 1996, pledges to the endowment fund and by way of operating and specific programme contributions from 54 governments and 115 other benefactors totalled US$ 318.6 million.[68] In addition to its Headquarters,

---

[68] The United Nations University (UNU). 1. The UNU Concept. 2. Organization. 3. Finance. In: UNU/INRA. UNU in Africa: Managing Natural Resources for

UNU has five research and training centres: the World Institute for Development Economics (UNU/WIDER) in Finland; the Institute for New Technologies (UNU/INTECH) in the Netherlands; the International Institute for Software Technology (UNU/IIST) in Macau; the Institute for Natural Resources in Africa (UNU/INRA) in Ghana; and the Institute for Advanced Studies (UNU/IAS) in Japan. The University also has three research and training programmes: the International Network on Water, Environment and Health (UNU/INWEH) in Canada; the International Leadership Academy (UNU/ILA) in Jordan; and the Programme for Biotechnology in Latin America and the Caribbean (UNU/BIOLAC) in Venezuela.[69]

In 1996 alone, UNU published 22 books and a great number of articles in journals, working papers, discussion papers, and research reports in fields such as

- peace and governance (peace and security; democracy and human rights; governance, multilateralism and leadership),
- development (globalization, liberalization, and development; distribution, development and the economics of transition; the role of technology policy in industrialization and industrial competitiveness; technological change, and economic and social exclusion; mega-cities and urban development),
- environment (sustainable resource management; eco-restructuring for sustainable development; natural resources in Africa; water, environment and health),
- science and technology (national systems of innovation, science and technology institutions; software technology for developing countries; applications of biotechnology for development; microprocessors and informatics; food and nutrition; science and technology for human needs).[70]
- A special UNU report was devoted to an evaluation of UNCED's Agenda 21, and UNU's contribution to its

---

Sustainable Development. The United Nations University. Institute for Natural Resources in Africa (UNU/INRA), Tokyo, April 1997, pages 27/28.
[69] Five Years after Rio: UNU's Responses to Agenda 21, The United Nations University, Tokyo, 1997, Footnote 5 on page 36.
[70] The United Nations University Annual Report 1996.

implementation.[71] It is subdivided into 33 chapters under three general headings: Eco-restructuring, Managing natural resources, Environmental governance.

The booklets "New Titles", issued bi-annually by the United Nations University Press, contain impressive lists and descriptions of UNU publications on many aspects and details of the global problems in the fields mentioned above, and produced "by scholars in the UNU global network".

2.4 The United Nations Educational, Scientific and Cultural Organization (UNESCO)

For historical reasons, it may be appropriate to mention under this heading the World Conservation Union, IUCN (see also 2.9.1). "IUCN is a unique Union of more than 880 state, government agency and NGO members in 133 countries, as well as a volunteer network of more than 8,000 internationally acknowledged scientific, technical, and legal experts contributing to conservation through IUCN's six global commissions. Founded in 1948, IUCN has offices in over 40 nations and is headquartered in Switzerland. IUCN's mission is to influence, encourage and assist societies throughout the world to conserve the integrity and diversity of nature and to ensure that any use of natural resources is equitable and ecologically sustainable".[72] Fifty years ago, UNESCO was instrumental in the creation of IUCN, as Prof. Federico Mayor, the Director-General of UNESCO, reminded his audience in his address at the Special Session of the UN General Assembly on June 25, 1997. Federico Mayor continued: "Since that time it (UNESCO) has played a leading role, with its UN partners and the scientific community, in **the development of international scientific programmes that address environmental and developmental problems in an integrated manner**. I would mention here in particular the Intergovernmental Oceanographic Commission - with its important work on climate change and ocean health, leading the development of the Global Ocean Observing System (GOOS) -

---

[71] See footnote 69.
[72] Considerations For The United Nations General Assembly Special Session to Review and Appraise the Implementation of Agenda 21, IUCN Headquarters, Gland, June 15, 1997.

and UNESCO's international scientific programmes concerned with fresh water and with Man and the Biosphere. Several new interdisciplinary initiatives have been developed (by UNESCO) in response to UNCED, notably the World Solar Programme 1996-2005 promoting all forms of renewable energy ...".[73] (Bold type by K.G.)

In Venice, UNESCO has a Regional Office for Science and Technology for Europe (ROSTE). Its activities include

- basic sciences, with special attention to environmental and health problems;
- engineering and applied research, including research on energy and new materials, and on environmental technologies;
- ecology, hydrology, oceanography, coastal zone management;
- the transformation of scientific communities in Europe;
- cultural projects.

For the execution of its projects, ROSTE uses the principle of networks, subdivided as follows:

- Promoting research;
- Organization of training activities;
- Improving information exchanges;
- International independent expertise exercises ("Peer Review");
- Promoting mobility of scientists by offering opportunities for research and sponsoring their participation in scientific meetings and training courses;
- Organizing high level colloquiums stimulating debates on principal scientific problems in Europe;
- Fund raising for activities of particular importance within the objectives of UNESCO.[74]

In the context of this programme, ROSTE organized, for instance, the "Genoa Forum of UNESCO on Science and Society" which issued the "Genoa

---

[73] UNESCO, Address by Mr. Federico Mayor, Director-General of UNESCO, at the Special Session of the General Assembly, Earth Summit + 5, New York, 25 June 1997.
[74] NEWS FROM UNESCO IN VENICE, Issue No.4, July - December 1995.

Declaration on Science and Society".[75] Another example is the Advisory Panel on Science Policy to the Ministry of Science of the Russian Federation which ROSTE organized at the request of the Ministry, and which met with Ministry officials in Moscow in 1993. The present author was a member of both the Genoa Forum and the Moscow Panel.

In order to obtain advice on UNESCO's Medium Term Plan, the Director-General of UNESCO had set up an Advisory Committee on Science, Technology and Society of which the present author was also a member (1981-1983).

2.5   The World Health Organisation (WHO)

Another UN Agency regularly receiving scientific advice is the World Health Organization (WHO). In its capacity as the only global intergovernmental health agency, WHO provides a framework for support of international health activities by Member States. Some international activities in areas such as child welfare, nutrition and education, promoted by UNICEF, FAO and UNESCO, are closely coordinated with those of WHO.[76] The six principal objectives for the WHO activities are as follows:

- development of comprehensive health services
- disease prevention and control
- promotion of environmental health
- health manpower development
- promotion and development of biomedical and health services research
- programme development and support

An **Advisory Committee on Medical Research (ACMR),** set up in 1959, meets annually to advise the Director-General of WHO on development of the

---

[75] GENOA FORUM OF UNESCO ON SCIENCE AND SOCIETY, UNESCO Venice Office, 1996.

[76] The Role and Activities of WHO in the Field of Health Related Research, Short summary statement for distribution to delegations participating in the "Scientific Forum" of the Conference on Security and Co-operation in Europe, WHO Regional Office for Europe, 1980.

Organization's global research programme. Regional advisory committees were also established to deal with matters of regional importance.

All WHO programmes include research components monitored by advisory groups. Recommended research activities are implemented at national level through appropriate networks of collaborating and participating centres and bodies (research institutes, universities, medical research councils and medical academies).

2.6 The United Nations Development Programme (UNDP)

UNDP is the central coordinating body for technical co-operation in the UN system. It has a number of associated funds and programmes, each funded separately through voluntary contributions, which provide specific services through the UNDP network. For instance, the UN Fund for Science and Technology for Development (UNFSTD) focuses on endogenous capacity-building in science and technology, information and quality control; and technology assessment, innovation and entrepreneurship. The Fund also deals with energy conservation and new sources of energy. With the assistance of **scientific advisers from outside**, UNDP produced a Human Development Report 1994, already mentioned in section 1, which contains a wealth of useful data. Apart from UNDP's international staff of 1,045 and national staff of 5,714 working in 136 UNDP country and liaison offices worldwide, 1,723 UN Volunteer specialists, 6,203 international experts, and 17,181 national experts served on UNDP programmes and projects in 1994.[77] Under a UNDP programme, the present author served as consultant to the Academy of Sciences of the Republic of Moldova on co-operation with Western countries.

2.7   The United Nations Environment Programme (UNEP)

Twenty-five years ago, meeting in Stockholm at the United Nations Conference on the Human Environment, governments established the United Nations Environment Programme (UNEP). "Throughout its history, UNEP has actively promoted environmentally sound development, which seeks to maintain economic progress without damaging the environment and the

---

[77] BUILDING A NEW UNDP, UNDP 1994/1995 Annual Report.

natural resource base upon which future development depends. UNEP has served as an **expert 'watchdog', monitoring the state of ecosystems and species worldwide**. It has been, and remains, the environmental conscience of the United Nations. UNEP has played an instrumental role in the adoption of international environmental conventions and treaties aimed at preserving the ozone layer, conserving biological diversity, coping with climate change, protecting the oceans and seas, controlling the movement of toxic wastes and controlling the trade in endangered wildlife species. ... As UNEP and its partners continue to accumulate **scientific knowledge on the adverse environmental impacts of human activities, additional issues emerge that demand the international community's urgent action**. As we approach the new millennium, **we face a range of complex, long-term environmental problems** portending immense consequences for the economic well-being and security of nations throughout the world: global warming, depletion of the ozone layer, the decline of biodiversity, the loss of soil and forests, contamination of our fresh water supplies, vanishing fisheries and the flood of toxic substances entering our environment and our bodies, threatening our physical and reproductive health. Recognizing that **UNEP needs the appropriate tools and resources** if it is to tackle these issues, governments meeting at the 19th session of UNEP's Governing Council ... adopted the Nairobi Declaration, giving the Programme a revitalized and strengthened mandate. ..."[78] (Bold type by K.G.)

2.8 United Nations Institute for Disarmament Research (UNIDIR)

"UNIDIR is an autonomous institution within the framework of the United Nations. It was established by the General Assembly for the purpose of undertaking independent research on disarmament and related problems, particularly international security issues. ...

UNIDIR has a very small permanent staff. It relies predominantly on project-related short-term contracts to implement its research programme. This system of recruitment permits recourse to and **utilization of reputable expertise** available both inside and outside the UN system. In addition, this mode of functioning contributes to the efforts of the Institute to cooperate with other institutions and individual experts as a means of sharing their knowledge.

---

[78] Kofi Annan, An Indispensable Contribution, Our Planet, Vol.9, No 1, June 1997.

The research projects of UNIDIR are carried out within the Institute or are commissioned to individual experts or research organizations. In the latter case, UNIDIR approaches those whom it considers qualified to be engaged in the respective projects. It determines the framework of research and subsequently reviews it before completion and publication. **For some major studies, multinational groups composed of persons known for their expertise and experience are established so that a multidisciplinary approach and various schools of thought may be taken into account. ...**"[79] (Bold type by K.G.)

2.9 Non-Governmental Organisations (NGOs)

The examples given in the preceding sections 2.1 - 2.8 were chosen under the aspect that particular units, sub-organizations, programmes, projects and conferences of the UN system needed scientific and scholarly advice and, in order to obtain it, established an advisory mechanism that, hopefully, served their needs. In this section we are going to look at some non-governmental scientific organizations (NGOs) which, at their own initiative and under their own responsibility, study problems of general concern and offer to governments, the United Nations, the public, and whoever is interested, their expert advice with respect to the available solutions.

"NGOs can have functions of various kinds:

- They may be active in problematic fields which governments have not yet taken up adequately, or not at all.
- They may have been founded in order to keep the government out of certain fields; the government may then restrict itself to just controlling the results.
- They may exert functions of control and correction with respect to governmental claims of direction.
- They establish ·themselves as important "players" in the world community; groups active in development policy, human rights and environmental organizations contribute towards the formation of an international public.

---

[79] United Nations Institute for Disarmament Research. The Institute and its Activities, Palais des Nations, Geneva

About 2,000 NGOs have obtained observer status by the United Nations. In world conferences they appear as new actors of international politics".[80]

Again, a few examples will be given.

2.9.1 The Expert Panel on Trade and Sustainable Development (EPTSD) of the World Wide Fund for Nature (World Wildlife Fund, WWF)

WWF being closely related to IUCN (see section 2.4), EPTSD set itself the goal to initiate the design of innovative policy packages and integrated policy instruments that will maximize synergies and aim to minimize conflicts between trade, environment and development objectives. In its first meeting which EPTSD held in Glion, Switzerland, from October 28-30, 1996, it was agreed to undertake sectoral examination of specific commodities/sectors to illustrate problems of environmentally unsound processing and production methods and generate broader conclusions from them. Timber, textiles, and electricity generation were chosen as first examples. Three main tasks were defined for EPTSD, namely:

- to establish whether or not the selected case studies can feed into generic recommendations on the formulation of policy packages which harness trade to sustainable development,
- to consider whether and how these sectoral studies, and the work of the panel generally, can contribute to broader public understanding of trade and sustainable development issues,
- from a consideration of the effects of national and multilateral policies on market access and processing and production methods choices, and consequently on sustainable development, to further develop policy formulation approaches to developing packages of integrated policies which maximize the positive contribution of trade to sustainable development.[81]

---

[80] Dirk Messner, Franz Nuscheler, Global Governance, Policy Paper 2, Stiftung Entwicklung und Frieden, April 1996. (English translation by K.G.)

[81] Expert Panel on Trade and Sustainable Development, 2nd Meeting, Cairo, 16-18 February 1997. EPTSD Secretariat Report, WWF, 1997.

2.9.2 The International Council of Scientific Unions (ICSU)

"More than a century ago, scientists from different nations and disciplines identified the need for a forum in which they could work together on common concerns. Soon after World War I , several national organisations established an International Research Council, which became the International Council of Scientific Unions (ICSU) in 1931. Today, ICSU membership includes 23 international Scientific Unions, each with its own constituency, traditions, and objectives ... ICSU also has 94 National Scientific Members that represent the scientists in individual countries, typically through their academies or research councils. Altogether, through these disciplinary and national bodies, more than 130 nations participate in ICSU's work.

To design and implement international interdisciplinary scientific programmes is one of the goals of ICSU. **Members of ICSU have established a number of interdisciplinary bodies to address specific problems that require international collaboration. Through these bodies, ICSU provides a unique source of independent and scientifically authoritative analysis**. Since such assessments and guidance are free from the biases and constraints of national interests, they are generally regarded as the world's best available consensus of fact and scientific judgement. (Bold type by K.G.)

Very many of the constituent bodies of ICSU are concerned with environmental issues ... ICSU has an Advisory Committee on the Environment (ACE) to advise its Executive Board on how best to coordinate this complex range of environmental activities.

ICSU has been closely involved in preparatory and follow-up work relating to the UN Conference on Environment and Development (UNCED). Before the meeting, held in Rio in 1992, ICSU together with other bodies held an international conference to develop an Agenda of Science for Environment and Development into the 21st Century (ASCEND 21). This report, and the subsequent participation of many ICSU scientists at UNCED, helped shape Agenda 21 - approved at Rio as the scientific blueprint for the planet's future, providing important guidelines for scientists and policy makers alike. ...

As ICSU's environmental programmes grow in number and level of activity, relationships with governments, industry and commerce - that ultimately provide the necessary resources - become ever more important. ICSU is therefore strengthening its links with the growing environmental activities of intergovernmental organisations, such as the Organisation for Economic Co-operation and Development (OECD) and the World Bank. ...

The Scientific Committee on Problems of the Environment (SCOPE) is ICSU's pioneer activity relating to the environment. Established in 1969, SCOPE originated as an interdisciplinary body of natural science expertise which addresses constraints of society on the environment as well as the human response to environmental issues. ...

The need to understand our planet has motivated scientists to develop worldwide linkages and working partnerships. Four major research programmes, organised through ICSU and its partners, provide a framework for these efforts. They identify scientific priorities and develop standard methodologies that are widely used in national programmes, thereby helping to foster and integrate research collaborations. The results of these programmes help to provide policy makers with the best available scientific knowledge for setting strategies for sustainable development.

The following aspects of Earth system science are covered by these programmes:

- the physical climate system, comprising the dynamics of the atmosphere, the oceans, the land surface and ice sheets (**World Climate Research Programme, WCRP**)
- the global interactions between living and non-living processes, that together underpin the habitability and productivity of our planet (**International Geosphere-Biosphere Programme, IGBP**)
- the structure and function of biological diversity; covering plant, animal and microbial life, on land, in freshwater and at sea (**DIVERSITAS**)
- the interactions between human society and its environment on a planetary scale (the **International Human Dimensions Programme on Global Environmental Change, IHDP**). ..."[82]

---

[82] Understanding Our Planet. An overview of the major scientific activities of ICSU and its partners that address global environmental change. Second Edition. International Council of Scientific Unions, Paris 1996.

2.9.3   The InterAcademy Panel on International Issues (IAP)

"IAP was created in 1995 as a forum of the world's scientific academies, to work together in providing advice and input to governments and international organizations and in informing public opinion on scientific aspects of issues of concern internationally. Through bilateral, regional, and worldwide collaboration, IAP contributes to building the capacity of academies to contribute to meeting major challenges faced collectively or individually by our countries. To date, over 70 academies, representing countries from all regions of the world, have participated in IAP-organized projects.

The impetus for creation of IAP grew out of concern among scientific academies for the consequences of world population growth. ... In 1992, The Royal Society, the Royal Swedish Academy of Sciences, the Indian National Science Academy, and the US National Academy of Sciences proposed a summit of academies to address the population issue. They were joined by an additional 12 academies in sponsoring the summit. In October 1993, representatives of the world's scientific academies met together for the first time in New Delhi, India, in a summit of academies, to address the issue of growing world population.

At the New Delhi population conference, sixty academies signed a Population Statement reflecting their continued concern about the problems of population growth, resource consumption, and the environment. This population statement was forwarded to governments and international decision makers, especially those at the 1994 UN International Conference on Population and Development in Cairo.

The day following the 1993 Population Summit, 16 academy representatives met to review the success of the meeting and to discuss next steps. It was decided to create a framework for co-operation among scientific academies, the organizational arrangements to be determined at a later date.

In January 1995, ... 14 academy representatives convened the first meeting of a Steering Committee to agree upon the organization of a cooperative framework among academies. At that meeting it was decided to broaden the focus of possible activities beyond the issue of population. Accordingly the name of the organization would be the InterAcademy Panel on International Issues.

As envisioned by the initial organizers, the IAP would serve as a **forum through which academies worldwide can use their specific expertise to bring together leading authorities in the natural and social sciences, in**

order to advise governments and international organizations and to inform public opinion on scientific aspects of issues of concern internationally or of concern to multiple academies.

Membership in IAP is limited to one academy of natural science per country, with preference being given to academies that were involved in international activities (e.g. adherence to ICSU) and were constitutionally able to engage in advice giving activities. Any academy meeting these criteria and wishing to participate in IAP would be accepted.

The IAP operates via a steering committee whose work is facilitated by two co-chairs, one from a developing country academy and the other from a developed country academy. ... The steering committee is supported by a secretariat, serving for a three-year term, housed at a volunteering academy. The Royal Society of London is serving as the initial secretariat. ... The IAP is developing a communications network among academies including a public newsletter and electronic linkages. ...

Among the initial activities of the InterAcademy Panel was participation in the United Nations Second Conference on Human Settlements (Habitat II) ...

FUTURE IAP PROJECTS

... The Steering Committee agreed to pursue the following activities:

Population Summit Follow-up. ...

Year 2000 Conference on Transition to Sustainability.

The Steering Committee agreed to begin preparations for a Year 2000 IAP conference of scientific academies to address opportunities and challenges for a worldwide transition to demographic, economic, and environmental sustainability in the 21st century. **A product of the conference would be a consensus statement, jointly developed by the world's scientific academies, on the challenges to sustainable development in the coming century and ways that science and technology can contribute to meeting those challenges. ...**[83] (Bold type by K.G.)

---

[83] John P. Boright, The InterAcademy Panel and Plans for a Year 2000 Conference on the Sustainability Transition, IX International Amaldi Conference of Academies of Sciences and National Scientific Societies, SECURITY QUESTIONS AT THE END OF THE TWENTIETH CENTURY, Geneva, 21-23 November 1996, United Nations and CERN. REPORT AND DOCUMENTATION. Accademia Nazionale dei Lincei, Rome 1997.

2.9.4   The International Institute for Applied Systems Analysis (IIASA)

IIASA was founded in 1972 during the Cold War at the initiative of the USA and the USSR, in order to reduce the tensions between the two opposing blocs by scientific co-operation on problems of general interest which urgently required advice-giving to national policy makers and the world public. Many European countries, East and West, as well as Canada and Japan joined, among them the Federal Republic of Germany and the German Democratic Republic (GDR) at a time when the GDR was not yet recognized as a sovereign state by the Western powers. By 1975 project groups for the following research areas had been set up:

- Biological and medical systems
- Computer science
- Ecological and Environmental systems
- Energy systems
- Integrated industrial systems
- Methodological problems of applied systems analysis
- Design and management of large organisations
- Water resources
- Management of urban and regional systems

Today, IIASA is supported by a group of member organisations in 17 nations (Austria, Bulgaria, Canada, Czech Republic, Finland, Germany, Hungary, Japan, Kazakhstan, Netherlands, Norway, Poland, Russia, Slovak Republic, Sweden, Ukraine, USA) and continues to contribute to the promotion of knowledge in important fields such as energy technology and energy modelling, dynamics of ecological systems, sustainable development of Russian forests, transition economies, climate impact, nuclear safety and land use, to give only a few examples. IIASA would be able to make important contributions to a new scientific advisory system for the United Nations.

2.9.5 The Pugwash Conferences on Science and World Affairs

In this context the *Pugwash Conferences on Science and World Affairs* should also be mentioned. "They deal with the dangerous problems deriving from the existence of weapons of mass destruction, and with the options for getting rid of these threats to the survival of civilized humanity and to the health of future generations. Their purpose is to bring together, from around the world, influential scholars and public figures concerned with reducing the danger of armed conflict and seeking cooperative solutions for global problems. Meeting in private as individuals, rather than as representatives of governments and institutions, Pugwash participants exchange views and explore alternative approaches to arms control and tension reduction with a combination of candour, continuity, and flexibility seldom attained in official East-West and North-South discussions and negotiations. Yet, because of the stature of many of the Pugwash participants in their own countries (as, for example, science and arms-control advisers to governments, key figures in academies of science and universities, and former and future holders of high government office), insights from Pugwash discussions tend to penetrate quickly to the appropriate levels of official policy-making".[84] By the end of 1996 there had been well over 200 Pugwash Conferences, Symposia, and Workshops, with a total attendance of some 10,000. (There are now in the world over 3,000 "Pugwashites", namely individuals who have attended a Pugwash meeting). The first Pugwash Conference was held in 1957 in the village of Pugwash, Nova Scotia, which gave its name to all the following conferences. The first part of Pugwash's history coincided with some of the most frigid years of the Cold War, marked by the Berlin Crisis, the Cuban Missile Crisis, the invasion of Czechoslovakia, and the Vietnam War. In this period of strained official relations and few unofficial channels, the fora and lines of communication provided by Pugwash played useful background roles in helping lay the groundwork for the Partial Test Ban Treaty of 1963, the Non-Proliferation Treaty of 1968, the Anti-Ballistic Missile Treaty of 1972, the Biological Weapons Convention of 1972, and the Chemical Weapons Convention of 1993. Subsequent trends of generally improving East-West relations and the emergence of a much wider array of unofficial channels of communication have somewhat reduced Pugwash's visibility while providing alternate pathways to similar ends, but Pugwash meetings have continued until the present to play an important role in bringing together key analysts and policy advisers for sustained, in-depth discussions of the crucial arms-control issues of the day: European nuclear forces, chemical and biological weaponry, space weapons, conventional force reductions and restructuring, and crisis control in the Third World, among others. Pugwash has, moreover, for many

---

[84] PUGWASH CONFERENCES ON SCIENCE AND WORLD AFFAIRS. (A Brief Description). Text issued by the Pugwash Central Office, London, September 1995.

years extended its remit to include problems of development and the environment[85].

As a Non-Governmental Organisation registered with the United Nations, Pugwash has raised its voice in UN fora. The worldwide network of experts built up by Pugwash over many years could be a valuable element in an institutionalized advisory mechanism for the UN.

3   Reasons why scientific advice is often ignored

We have seen in section 2 that there is a considerable number of institutions which, working independently and without overall co-ordination, have over the years built up their own advisory bodies and advisory methods which provided, and still provide, advice as to how certain goals might be reached, or how some specified disasters could be circumvented. (Some statements stressing the necessity of international and interdisciplinary collaboration for the treatment of global problems have been accentuated by bold type.) However, as is well known and as was described above, the situation of humanity and of its environment on this planet is not at all under satisfactory control. What went wrong? Was there no appropriate advice, available to governments and the public, on how to avoid the impasses in which we now find ourselves? Was the advice obtained of poor quality? Or was the necessary advice available, and the circumstances were such that policy-makers were unwilling or unable to accept the advice and the conclusions to be drawn from it?

There are several potential reasons why governments are often reluctant to accept advice offered by scientific experts.[86] We have listed them in chapter 6, section 3. There we have mentioned that, as a general rule, scientific advisers to governments should refrain from suggesting what should be done unless they are told precisely which goal is to be reached and which boundary conditions are to be observed in approaching this goal. What should be done is a political, not a scientific decision. But, of course, scientists must warn politicians and the public what might happen if certain measures are taken, or

---

[85] loc.cit. (Footnote 84).
[86] K.Gottstein, The Need for Neutral Scientific Advice in Complex Situations of High Risk, Proceedings of the 6th Amaldi Conference "A Contribution to Peace and International Security", Rome, 27-29 September 1993.

not taken. When scientists and scholars forecast what is going to happen they should clearly state under which assumptions they have come to that particular conclusion, and what the uncertainties are.

No doubt, many of the advisory bodies described above have followed this rule. The fact, however, that there are so many of them with different priorities, different points of view and different legitimizations - an advantage under the aspects of pluralism, scientific competition and peer review - resulted, nevertheless, in a splitting of impact, and facilitated criticism, in some cases perhaps justified, of one-dimensional thinking. In many cases the advice given did not take into account all aspects which successful politicians have to keep in mind, so that the latter felt entitled not to take serious what they were told by the experts of individual disciplines. There was no all-encompassing, interdisciplinary advice which coordinated the "sectoral" advice given by the experts of the various isolated disciplines.

The latter statement may be somewhat unjust with respect to institutions like ICSU, IAP, UNU, and IIASA, as described in section 2. They do indeed *aim* at providing the all-encompassing advice needed, at least for the future and within certain limits. The fact remains, however, that what is required in the way of timely interdisciplinary analysis of the available options for action, is more than these institutions are able to offer under present conditions. Therefore, instead of offering a survey of the different options available for coping with the situation, with the risks, benefits and costs estimated conservatively for each option, including the potential long-term consequences of side- and after-effects in other areas, they often limit themselves to isolated warnings and to recommendations as to what should, or should not, be done. In the absence of a complete picture in the light of *all* disciplines, it is then easy for politicians to evade the issue by concentrating on an area where the criticized measures have positive effects (for instance, by considering that disregard for environmental protection has short-term economic advantages).

This could be changed. With the assistance of science and technology it would no doubt be feasible in many cases also to foresee the occurrence of by- and after-effects of human activities. They could be taken into account and made part of the overall planning. Countermeasures could be prepared, or the activities planned could be replaced by others with less harmful consequences. As the main effects and the by-effects often occur in different fields of specialization (in security policy and in psychology, for instance, or in economy and in climate research), multidimensional thinking and interdisciplinary collaboration are required. Because of the global character of many problems, international co-operation is also indicated. In economic policy it is also necessary because laws in the field of environmental protection must be valid worldwide if they are to be effective. Environmental protection costs money. If a country would not enact laws which are valid in other countries, its industries would have an economic advantage. In order to maintain its economic competitiveness, no country would be prepared to set up effective environmental protection unless its competitors would be subject to similar laws.

4 The need for truly interdisciplinary advice to the United Nations

In section 1 of this chapter we have listed some of the pressing global problems of our time. These problems are interconnected in complicated ways. Attempts to solve one of them separately affect, often adversely, the solution of the other problems. What is needed, therefore, is a truly interdisciplinary and international approach. One nation alone, or one scientific/technological/ scholarly discipline alone, may be able to make useful contributions within a coordinated international and interdisciplinary approach but, if isolated, will only add to the flood of unrelated and unsorted information that confuses decision-makers, the media, and the general public.

In any case, the institutions of world governance, in order to be able to cope with exactly those problems which make world governance necessary, will be in need of lucid presentations of the roots and interconnections and different aspects of the problems in question, and of the available options for re-acting to them, with the costs, risks, and benefits estimated as accurately as possible for each option. There can be little doubt that a task of this complexity can only be tackled by permanent interdisciplinary and international Working Groups of the highest scholarly standard.

This requirement also applies, of course, to the preparation of peacekeeping operations by the UN itself, or by alliances acting on behalf of the UN. It is essential for the success of such operations that the parties involved in a conflict are approached with great sensitivity regarding their historical, ethnic, cultural and religious background.

Many examples could be given for the necessity of truly interdisciplinary study of global problems. A particularly good example is climate research.

Professor *Klaus Hasselmann,* Director of the Max Planck Institute for Meteorology, said this:[87]

"The development of an optimal, internationally agreed upon, strategy for climate protection is a very complex task of contemporary politics. Such a strategy would have to take into account the costs of adaptation (to a potential global rise of temperature) as well as those of avoiding changes of climate by restructuring industry. In order to support the efforts to fulfill this task, science should supply ground-data as reliable as possible. **This requires a close collaboration between climatologists, ecologists, economists, legal experts and representatives of other disciplines.** ... Scientifically unfounded 'opinions', whether by prophets of an approaching climate catastrophe, or by climate skeptics, do not bring any progress." (Bold Type by K.G.)

5   Suggestions for a new international and interdisciplinary advisory system for the institutions of world governance

We have seen in the previous chapters that many of the urgent problems with which humanity is confronted today, require policy decisions taken at the global level, and implemented globally. These decisions must be based on sound knowledge of the facts, and of the potential by- and after-effects connected with each option available for political action. So far, there is no substitute for the UN, as far as decision-making at the global level is concerned. In fact, as mentioned in section 1, there are efforts underway to reform and strengthen the UN in order to make it better prepared for its task of "world governance". This means that the UN will also need a better advisory system. We have seen in section 2 that the various advisory bodies existing in different branches of the UN system, while useful, do not yet fulfill the requirements of an overall network of scientific analysis that would be able to act as an efficient warning system against impending disasters.

There can be no doubt that the world community will be in need of reliable, interdisciplinary scientific advice if major disasters are to be avoided. Because of the complexity of this task and the considerable cost involved in setting up a network of highly competent interdisciplinary working groups for this purpose, an extraordinary effort will be required. This should be the responsibility of the international community of the institutions of science, humanities and international law. There is no other international community or

---

[87] MPG-Spiegel 2/1996, p. 31-33. (Translation by K.G.)

institution with comparable access to unbiased expertise, and at the same time with sufficient prestige to make itself heard. The U.S. National Academy of Sciences (USNAS) with its effective Research Council and its Standing Committees on urgent issues could be an example on a national level of what will soon be required on an international scale. The USNAS, at the request of the U.S. Government or on its own initiative, has, for many years, taken up questions for the solution of which scientific advice is obviously required, and set up committees for their analysis. It maintains, as permanent institutions, a Committee on International Security and Arms Control (CISAC) and a Committee on Science, Engineering and Public Policy (COSEPUP).

It had been the suggestion of the present author that the international community of national academies of sciences and national scientific societies could fill this gap, and that this might be organized under the auspices of the Amaldi Conferences which are already established internationally as a joint enterprise of this community, dealing in a fruitful manner with the global problem of nuclear disarmament and, in their "window sessions", with some other global problems.[88] It seemed possible to extend their agenda, with different experts and in parallel meetings, to include surveys of the obstacles to the solution of other global problems as well.

In the meantime, however, IAP was founded as a joint undertaking of academies and scientific societies (see section 2.9.3), and there is hope that it will develop into the advisory body that is so badly needed. Moreover, there is ICSU and the other institutions mentioned in section 2. There are also other candidates, like the European Science Foundation with its joint research programmes, among them one on Environment and Health.

---

[88] K.Gottstein, The Role of National Academies,Proceedings of the 4th Amaldi Conference, Cambridge 1991.

K.Gottstein, International Security in a Wider Sense, and the Amaldi Conferences, Proceedings of the 5th Amaldi Conference, Heidelberg 1992.

K.Gottstein, The Need for Neutral Scientific Advice in Complex Situations of High Risk, Proceedings of the 6th Amaldi Conference, Rome 1993 .

K.Gottstein, Collaboration of Academies and the Future of the Amaldi Conferences, Proceedings of the 8th Amaldi Conference, Piacenza 1995.

K.Gottstein, The Role of National Academies and Scientific Societies in Supplying Advice on the Nature of Global Problems and on the Available Options for Coping with Them. Introductory Remarks, Proceedings of the 9th Amaldi Conference, Geneva 1996.

Nevertheless, progress is slow, the meetings of the institutions mentioned are few and with long periods of time in between, and the transfer of their results into the political process is quite insufficient for rational decision-making. Meanwhile, the deterioration of the global situation continues rapidly with the immense growth in world population and in world production, severe damages to the environment and to biodiversity, a growing economic gap between the affluent and the poor, the dangers of civil war, the problems of "globalisation". There is no widely accepted theory as to where these developments are going to lead us, and what remedies might be available. Politicians are helpless, they are just trying to "muddle through". It is obvious that we are at the doorstep of a new world system for which an adequate global mechanism of checks and balances has not yet been found. This means that humanity is facing grave dangers. We cannot afford to sit waiting. At least, the available options for the avoidance of major catastrophes should be studied systematically and continuously. The results should be communicated to the Secretary-General of the United Nations who should be advised to urge governments to conclude agreements on the necessary actions. The implementation of such agreements should be continuously monitored by Standing Committees.

As there is no longer, since the demise of ACAST (see 2.1), a central scientific advisory system for the UN, it should be considered whether the international community of academies of sciences and of national scientific societies should not seek to be asked by the Secretary-General of the UN to set this up for him. This would require an extraordinary, but worthwhile, effort by the academies and scientific societies to develop, on the foundations laid by IAP, ICSU, and other organizations, an effective system of permanent working groups for the interdisciplinary study of policy options, with their probable by- and after-effects, for tackling the well-known population, nutrition, environment, migration, technology, economic, social, political and othr problems of global relevance.

The Amaldi Conferences, in their annual "window sessions", could present a survey of the progress made in the preceding year in the creation of such a global scientific advisory system for the United Nations.

Acknowledgement

I am indebted to *Dr. Fatai Kayode Salau* for some new literature on the United Nations University, the World Conservation Union, and EPTSD (see chapter 2) which he obtained at the "Earth Summit Plus Five", the special session of the UN General Assembly, held in New York in June, 1997.

21 The Scientific Culture and Its Role in International Negotiations [89]

1 Historical introduction

Past history offers examples of scientists acting as diplomatic negotiators. The great philosopher, mathematician, engineer and geologist *Gottfried Wilhelm Leibniz* (1646-1716) served as envoy of his prince elector, the Archbishop of Mainz, to King Louis XIV of France whom he tried to persuade to attack Egypt rather than the German Holy Roman Empire. Later he travelled all over Europe, negotiating with the Emperor in Vienna and Tsar Peter the Great, being instrumental, by his historical research and his personal connections, in securing for the House of Hanover first the electorship of the Holy Roman Empire and later the succession to the English throne. He also tried actively in many negotiations to re-unite the Christian Church.

Another well-known example is *Benjamin Franklin* (1706 – 1790) the scientist and diplomat. He did research in electricity, invented the lightning rod, was American envoy to England and France where he negotiated the treaty by which France agreed to support the American revolutionaries.

Let us now jump to the 20th century. During the Second World War scientists served as government advisers and scientific diplomats. In 1940 a National Defense Research Committee was created in the United States. One year later it was merged with the Office of Scientific Research and Development (OSRD) within the Executive Office of the President. Its head, *Dr. Vannevar Bush*, had direct access to President Roosevelt and considerable influence within the government bureaucracy. A branch of OSRD was set up in London, whereas Great Britain opened in Washington the British Central Scientific Office (BCSO). Later Scientific Liaison Offices of Australia, New Zealand, Canada and South Africa were added, forming with BCSO the British Commonwealth Scientific Office. Its task was the co-

---

[89] Contribution to the PIN-Project ("The Processes of International Negotiation"), to be published in: Professional Cultures and International Negotiations: Problems and Opportunities.

ordination of the co-operation with the United States in developing the atomic bomb, radar and other systems, for instance for the warfare against submarines, in which modern science and technology played a leading part. The first directors of this Office were internationally known scientists: *Sir John Cockcroft*, *Sir Charles Darwin*, *Sir Thomas Taylor*. Later, the head of the Office received the title and responsibility of "Attaché for Scientific Questions" to the British Embassy in Washington. This institution proved so useful that it was retained after the end of the war.

In the late 1940s and in the 1950s the United States sent university professors as Science Attachés to several of its European embassies. They were, however, almost exclusively representatives of American science, without much contact to the policy-makers of their country. The "Sputnik Shock" of 1957 changed this situation. The State Department set up a large division of science and technology, with subdivisions for atomic energy, space technology, environmental protection and general science policy. A new science attaché program was started. By 1970 the United States had science attachés in 23 countries. Other countries followed the example of the United States and Great Britain. At the same time about 25 embassies in Washington had a science attaché on their staff. The fields in which these science attachés were active, by observing developments in the United States and negotiating, under directives from their governments, agreements on co-operation with the United States, included:

- *arms control and disarmament* (but not weapons development which was under the jurisdiction of the Military Attaché)
- *energy and material resources*
- *industrial productivity and competitiveness* (as far as it depends on technology)
- *information on scientific developments in their home countries for interested circles in the U.S.*
- *education in science and technology*
- *scientific solutions to social problems (environmental protection, public transport, urban development, reactor safety, health care, crime prevention, data processing, etc.)*
- *science and technology policy*

Obviously, science attachés have to be in close contact with all institutions in their host countries which are engaged in these fields, in government as well as in parliament (Congress), in scientific institutions, academies, societies as well as in private industry. Of particular interest is the way in which these different bodies co-operate or, at least, interact when specific goals are to be reached.

2 Negotiations in an Interconnected World

Our modern world is so intertwined and interlinked in many ways that linear, monocausal thinking and acting is no longer adequate in an increasing number of cases. The world, of course, has always been a "world-wide web" but only in our scientific-technological age has it become necessary for human policies to take that into account. Due to the interconnections provided by modern means of communication and transportation and due to the many interfaces between education, economy, ecology, social conditions, foreign relations, national and international security, no single discipline of knowledge is able to assess the after-effects and by-effects that any policy measure taken in one field might have in all the other fields. In many instances, interdisciplinary and international co-operation will be required for a satisfactory solution of the problems at hand.

Important events that occur on the globe in one spot are immediately known everywhere. Repercussions of disturbances in one part of the world are felt over long distances in other parts as well. Television, computer networks, jet airplanes have made distances shrink. The emotions created by modern media with their reports on natural and political catastrophes in (formerly) distant lands, the arrival at our doorstep of the refugees from these catastrophes have deeply influenced the political life in modern democracies. The ordinary citizen, in his living-room, can be "present" when world events happen. This has changed drastically the role of diplomats. While stationed in foreign countries and representing their governments in bilateral and multilateral negotiations, they are now in constant and immediate contact with their superiors at home whose guidance and advice is available to them at all places and all times, without delay.

But also the subjects and topics of international negotiations at the end of the 20th century are increasingly related to science: Questions of arms control, of the verification of non-production or of the disposal of nuclear, chemical and biological weapons, of nuclear non-proliferation, environmental protection, food and agriculture, water supply, health, energy, population policy, etc., have strong scientific components. Consequently, the presence and participation of scientific experts is often required in international negotiations. Even in the more classical fields of conflict prevention and crisis management, which used to be the exclusive domain of experienced diplomats, there exist now academic experts in political science and international relations who specialize in peace and conflict research and, in particular, in the psychology of ethnic and international conflicts, and who can sometimes offer useful advice to the official negotiators.

The knowledge required for the successful settlement, in international negotiations, of the political problems of our time is often of an interdisciplinary nature, as indicated above. Therefore, it may not be sufficient to have scientific experts of one academic field as advisers, for instance physicists. Representatives of other disciplines of knowledge, for instance biologists, chemists, physicians, historians, psychologists, might be required as well if a satisfactory solution to the negotiated problem is to be reached.

Usually, there exist several possible solutions, each with its specific advantages, disadvantages, risks and costs. Which of these solutions seems to be preferable, depends on the interests of the parties negotiating with each other. An obvious question is: Are the benefits, risks and costs associated with any one solution distributed equally among the negotiating parties and/or the governments and nations they represent, or will one party bear most of the burdens while another one will enjoy most of the benefits? Decision-makers aware of their long-range responsibilities should also ask: Will only the present generation have the advantages from the chosen solution whereas future generations will have to pay the costs?

3   Interdisciplinary analysis of available options

In order to allow carefully balanced answers to questions of this type, a thorough analysis of the relevant facts and developments, their interconnections and trends, and of the probabilities and uncertainties involved, will be necessary. In many cases this will require lengthy, interdisciplinary investigations resulting in the preparation of various options. To choose among these options will be the task of the decision-makers in the centres of political power or of their deputies at the negotiation table. The scientists can only tell them what the options are, indicating the long-range risks, costs and benefits to be expected for each option. Given the complexity of many problems on the agenda of today's international negotiations, a considerable number of preparatory meetings of scientific experts is often required before clear-cut alternative options can be presented to the political decision-makers or their representatives, to be tested at the negotiation table. Some of these options, though technically feasible or even advisable, may not be realistic under the political, psychological or economic circumstances of the day. This will be known, however, to expert political scientists, psychologists and economists from their studies, and good politicians will feel it instinctively. Their experience will tell them.

Whenever a technical solution that seems necessary from an overall and long-range point of view, for instance in the area of environmental protection, turns out to be unacceptable politically, it will be necessary to determine carefully the nature of the obstacles. The next step should then be to work out soberly and unpolemically what the options are for overcoming the obstacles, and in which way the public could be informed and educated so that the removal of the obstacles becomes feasible politically.

Thus, in the scientific-technological world at the end of the 20th century, scientific and technological advice has become indispensable to successful politicians in the same way as rulers of earlier centuries recognized military, legal, and economic advice as useful, and often indispensable.

4 Scientists as negotiators

Sometimes, in international negotiations on scientific-technical subjects, the scientists themselves are the official negotiators for their countries. Usually, they are in this role in the preparatory meetings although also there some representatives of the foreign ministries, or of the ministries for research and technology, may be present and officially in charge. It is therefore of interest what the special features are that characterize scientists when they have to act as negotiators, as compared to negotiators from other backgrounds, such as diplomatic, military, legal, economic and other negotiators.

4.1 Special features characterizing scientists

The professional culture of scientists will be discussed in 4.3 but it may be appropriate already at this point to remind ourselves of some of the general advantages that scientists, as a group, enjoy:

- Scientists have studied science, and science represents a large portion of the knowledge of humankind.
- Scientists are used to dealing with complicated issues, to sorting and analyzing facts, drawing conclusions, and revising them if new facts make revisions necessary. In many cases, they will be able, in international and interdisciplinary co-operation, to work out options for political decisions, estimating the costs, risks and benefits for each option.
- Scientists can afford to think in terms of decades, to tackle long-range issues because, usually, they do not depend on re-election. (Politicians often understand the arguments just as well as scientists do. But often they believe that they cannot afford to follow the path of reason because that, they think, would make them lose the next elections.) So it is up to the scientists to inform the public of the long-range consequences of policies.
- Scientists are familiar with international co-operation. Many scientific problems can only be solved in this manner. This they have in common with many political problems. Co-operation creates ideas by exchange of views, it facilitates comparison of results, it stimulates criticism, it allows joint ventures and the realization of projects that would be beyond the scope of any one group or nation. It contributes towards an equilibrium of standards and forces, thereby reducing psychological and political tensions. International co-operation in science also opens up new channels of communication between different political systems.

Politicians, on the other hand, are forced to take short-term perspectives because

- they have to tackle the problems of to-day, they have to find the transition to the next day, in a similar manner as the "man or woman in the street" have to tackle his or her every-day problems,
- they must win the next election, as mentioned above, because otherwise they would lose their ability to influence the course of events. As long as the general public does not honour long-range thinking they feel justified in believing that they cannot afford it.

It must be said that scientists often do not have sufficient access to the realities of political life so that their advice sometimes disregards human and political factors and thus becomes unusable. This is why the analysis of the reasons for the non-implementation of scientifically convincing recommendations is so important, as indicated above. If necessary, the public must be convinced that the advice given on scientific grounds is sound, and that the politicians in charge should follow it if they do not want to lose the next elections.

A word of caution is in order here: Scientists are, of course, normal human beings. Usually, they love their work, are convinced of its importance, and want to bring it to fruition. In this sense they are biassed towards their own work. To counter-balance this inclination as far as possible, the institution of peer review has been introduced. The validity of scientific results, the importance of a scientific paper are not to be judged by the author himself but by some knowledgeable, usually anonymous, experts in the same field before a paper is accepted for publication, a project is funded, a prize is given, a professorship is offered, etc. In general, this works quite well. In some instances, however, it is difficult to find independent experts as judges or referees of this type. This may happen when a project is so specialized that the scientists working on it are the only experts available. This occurs sometimes in special projects of arms development and arms control, new reactor designs and their safety, the construction of large particle accelerators, etc. In a highly specialized field of this sort the number of experts may be limited to those who work in that field, know each other, and support each other. Anyone asked to give an independent expert opinion on a project in this field is either not independent or not expert.

4.2 Scientists as advisers to decision-makers

4.2.1 The U.S.National Academy of Sciences

To counteract this undesirable state of affairs, at least partially, the U.S. National Academy of Sciences (NAS) has set up Standing Committees on International Security and Arms Control (CISAC) and on Science, Engineering and Public Policy (COSEPUP). The members of these committees are prominent, independent scientists working in other fields than those under revision by their committees. However, with the support of the NAS these committees have established special working groups which study in depth the subject under investigation, thus gaining sufficient expertise for the publication of professional studies on projects requiring government decisions on funding and/or political action. A recent example is an investigation on the options for the disposal of plutonium from dismantled nuclear weapons[90]. These studies by NAS committees are generally recognized as being both expert and independent. CISAC, for example, has given itself three tasks:

- to inform members of the Academy on questions of security and arms control. Seminars are organized for this purpose,
- to create a cadre of independent experts on international security and arms control,
- to promote international communication and to offer advice to the U.S.Government on international security and arms control. In order to uphold political neutrality, a basic principle is: no agreements, no joint declarations, no publicity.

After NAS had set up CISAC in 1980, the USSR Academy of Sciences established a similar committee. The two committees met regularly twice a year, alternately in the U.S. and the Soviet Union. In an informal atmosphere, free from polemics, practical problems of security policy were discussed. The results were reported to the two governments. In parallel to the official arms control negotiations, these informal discussions of independent scientists

---

[90] National Academy of Sciences, Committee on International Security and Arms Control, *Management and Disposition of Excess Weapons Plutonium* (Washington, D.C.: National Academy Press, 1994).

contributed a great deal towards mutual understanding of motivations, goals, perceptions, difficulties, and preoccupations of both sides.

In September of 1986, the president of the NAS, *Dr. Frank Press*, suggested that European scientists, selected by European science institutions, should take part in discussions of this kind. After all, the security of the United States and the Soviet Union could not be separated from that of European countries on whose soil the nuclear weapons of the two blocs were deployed. Dr. Press invited the West European academies and scientific societies to consider forming a West European scientific committee on security questions which could be a partner of CISAC and its Soviet counterpart in their discussions of scientific security questions. A first exploratory meeting of CISAC members with European scientists interested in security questions had already taken place in Washington in June of 1986. It had positive results to which Dr. Press referred in his letter.

4.2.2 The Amaldi Conferences of Academies of Sciences and of National Scientific Societies

The Accademia Nazionale dei Lincei (ANL) in Rome was the first European science organisation to accept the American suggestion. It formed a "Gruppo di Lavoro per la Sicurezza Internazionale e il Controllo degli Armamenti" (SICA) and elected *Professor Edoardo Amaldi*, then Vice-President of ANL, as its chairman. Professor Amaldi organized a three-day "Workshop on International Security and Arms Control: The Role of the Scientific Academies" which took place in Rome in June of 1988. Invitations had gone to such members of West European academies and scientific societies and of the NAS who were assumed to be interested. The topics discussed at the workshop were:

- The treaty between the U.S. and the U.S.S.R. to eliminate intermediate range and shorter range nuclear missiles,
- The perspectives of drastic reductions in the strategic arsenals,
- The reconversion of weapon-grade fissionable material to peaceful uses,
- The future of the Strategic Defense Initiative (SDI). Points of view from Europe.

Each topic was introduced by a rapporteur and a co-rapporteur, with a general discussion following.

Because of the success of this meeting, Professor Amaldi, meanwhile president of the NAL, organized another conference of this type which took place in Rome in June of 1989. Whereas only U.S. and West European scientists had been invited to the first conference, this time members of the academies of the USSR, the German Democratic Republic, and of the other East European countries had also been invited and had accepted the invitation. Sixty scientists participated. The programme of the three-day meeting was divided into five sections:

- Deep cuts in nuclear weapons,
- Military stability in Europe: Prospects for reducing and restructuring nuclear and conventional forces,
- Conversion of weapon-grade fissionable materials,
- Prospects of a total ban of chemical and biological weapons,
- The role of academic institutions in the quest for peace and disarmament.

At the end of this meeting it was decided to repeat these conferences of academies of sciences and scientific societies on scientific questions of political relevance in annual succession. For 1990, Professor Amaldi offered again the hospitality of the ANL, but in December of 1989 he suddenly died. His successor as president of the ANL, *Professor Giorgio Salvini*, completed the preparations in Amaldi's sense, and opened the conference in Rome in June of 1990. This time, also the Chinese Academy of Sciences was represented. Scientists from 17 countries attended. The topics discussed concerned questions of international scientific and technological co-operation, particularly in environmental protection, in energy use, in climate research, in conversion of armaments production, in verification of technical disarmament measures, in the design of new security concepts, and in the relations between industrial and developing countries. From now on, in honour of the late Professor Edoardo Amaldi, these conferences were called **International Amaldi Conferences of Academies of Sciences and of National Scientific Societies**.

In the meantime, the Royal Society of London had formed an "ad hoc group on scientific aspects of international security" and the President of the Royal Society extended invitations to the next Amaldi Conference which took place in Cambridge in July of 1991 with 43 participants from Eastern and Western Europe, the United States and India. From then on the Amaldi Conferences finally did establish themselves, as previously decided, as annual events. Their venue was

- Heidelberg in 1992, with the Conference of German Academies of Sciences as host, supported by the Max Planck Society and the Deutsche Forschungsgemeinschaft,
- Rome again in 1993,
- Jablonna near Warsaw in 1994, with the Polish Academy of Sciences as host,
- Piacenza (the birthplace of Amaldi) in 1995,
- Geneva (under the auspices of the United Nations) in 1996.[91]

The Amaldi Conferences continue to discuss, from a scientific point of view, the current security situation. Recent topics were:

- the future of nuclear weapons,
- the technical and political aspects of the prevention of nuclear proliferation,
- the conversion of military R&D laboratories to civilian use,
- the control of conventional arms trade,
- the detection and removal of abandoned land-mines,
- the verification of the non-production and the destruction of chemical and biological weapons,

and related security questions. Usually, one session (called "window session") is devoted to a problem of "security in a wider sense". In this session, over the years, experts drew attention to, and discussed

- the human population explosion,

---

[91] Note added in December of 1998: In November of 1997 the Académie des Sciences hosted the 10th Amaldi Conference in Paris. In November of 1998 the 11th Amaldi Conference was held in Moscow at the invitation of the Russian Academy of Sciences. The 12th Amaldi Conference, in October of 1999, is to be held at Mainz, at the invitation of the Union of German Academies of Sciences and Humanities.

- the economic and social problems caused by the migration of millions of refugees from civil wars, persecution, hunger and natural catastrophes,
- the causes of nationalism and ethnic strife,
- the destruction of the ozone layer,

and others.

4.2.3 Global problems requiring negotiations based on scientific analysis

A constant subject of discussion at Amaldi Conferences in recent years was how the institutions of science could take a more active part in working out options for dealing with the urgent global problems. A tentative list of these problems, including those already mentioned, was already given in chapter 20:

- pollution of soil, water and air
- destruction of the ozone layer
- heating of the atmosphere
- desertification
- disappearance of animal and plant species in alarming numbers
- the human population explosion
- food and energy shortages
- migration of millions of people
- nationalism, racism and ethnic "cleansing"
- psychological, social, and economic instabilities
- civil wars and weapons trade
- the threat of nuclear proliferation and of the misuse of nuclear materials

These problems are interconnected in complicated ways. Attempts to solve one of them separately affect, often adversely, the solution of the other problems. What is needed, therefore, is a truly interdisciplinary and international approach on which the necessary international negotiations can be based. The national academies of sciences and the national scientific societies have, in principle, the tools for such an approach. They are interdisciplinary and they have close relations to their counterparts all over the world. The best experts in all fields of the natural and the social sciences - and in many cases also in the humanities - are among their members. Compared to politicians, to mention it again, these experts are rather independent because in general they have tenure and do not depend on re-election. Therefore, they can afford to concentrate on long-range goals rather than on short-range measures that may be popular at the moment but detrimental with respect to their long-range consequences.

4.2.4   ICSU, IAP, and ESF

Of course, there exist several scientific associations, projects and series of conferences which try to use the expertise and the connections of the scientific community for the design of rational and reasonable approaches to the tackling of some or all of the global problems which, in our days, threaten the peaceful development of humankind on this planet. The most universal of these associations are probably *The International Council of Scientific Unions (ICSU)* and *The Inter-Academy Panel on International Issues (IAP)*. *ICSU*'s activities include studies, projects, experiments and panels on ecosystem processes and biodiversity, on the relations between health and environmental pollution, on climate change, on world ocean circulation, on the global cycles of energy and water, on stratospheric processes, on global change and terrestrial ecosystems, on biodiversity and sustainable use of the Earth's biotic resources, to mention only some of them[92]. *IAP*, in close co-operation with ICSU, has similar objectives. IAP organized a Population Summit in New Delhi in 1993 and it contributed to the HABITAT II Conference of the United Nations held in Istanbul in June 1996. Other projects concerning sustainable development, resource consumption, energy needs and environmental protection are planned. The use of electronic communications for the exchange of ideas and for on-line conferences is envisaged[93]. ICSU and IAP are not the only international and interdisciplinary efforts by the scientific community to study global problems and inform governments and the public about them. There are also the joint

---

[92] For a survey of the activities of ICSU see Understanding Our Planet. An overview of the major scientific activities of ICSU and its partners that address global environmental change, 2nd edition, ICSU 1996.
[93] IAP Newsletter No. 1, March 1996.

research programmes of the *European Science Foundation (ESF),* among them one on Environment and Health.

4.2.5   The Pugwash Conferences on Science and World Affairs

In this context the *Pugwash Conferences on Science and World Affairs* should also be mentioned.[94] They deal with the dangerous problems deriving from the existence of weapons of mass destruction, and with the options for getting rid of these threats to the survival of civilized humanity and to the health of future generations. Their purpose is to bring together, from around the world, influential scholars and public figures concerned with reducing the danger of armed conflict and seeking cooperative solutions for global problems. Meeting in private as individuals, rather than as representatives of governments and institutions, Pugwash participants exchange views and explore alternative approaches to arms control and tension reduction with a combination of candour, continuity, and flexibility seldom attained in official East-West·and North-South discussions and negotiations. Yet, because of the stature of many of the Pugwash participants in their own countries (as, for example, science and arms-control advisers to governments, key figures in academies of science and universities, and former and future holders of high government office), insights from Pugwash discussions tend to penetrate quickly to the appropriate levels of official policy-making.[95] By the end of 1996 there had been well over 200 Pugwash Conferences, Symposia, and Workshops, with a total attendance of some 10,000. (There are now in the world over 3,000 "Pugwashites", namely individuals who have attended a Pugwash meeting). The first Pugwash Conference was held in 1957 in the village of Pugwash, Nova Scotia, which gave its name to all the following conferences. The first part of Pugwash's history coincided with some of the most frigid years of the Cold War, marked by the Berlin Crisis, the Cuban Missile Crisis, the invasion of Czechoslovakia, and the Vietnam War. In this period of strained official relations and few unofficial channels, the fora and lines of communication provided by Pugwash played useful background roles

---

[94] K. Gottstein, Die Pugwash-Konferenzen. Naturwissenschaftliche Rundschau 39, Heft 6, 253 - 255, 1986.
[95] PUGWASH CONFERENCES ON SCIENCE AND WORLD AFFAIRS. (A Brief Description). Text issued by the Pugwash Central Office, London, September 1995.

in helping lay the groundwork for the Partial Test Ban Treaty of 1963, the Non-Proliferation Treaty of 1968, the Anti-Ballistic Missile Treaty of 1972, the Biological Weapons Convention of 1972, and the Chemical Weapons Convention of 1993. Subsequent trends of generally improving East-West relations and the emergence of a much wider array of unofficial channels of communication have somewhat reduced Pugwash's visibility while providing alternate pathways to similar ends, but Pugwash meetings have continued until the present to play an important role in bringing together key analysts and policy advisers for sustained , in-depth discussions of the crucial arms-control issues of the day: European nuclear forces, chemical and biological weaponry, space weapons, conventional force reductions and restructuring, and crisis control in the Third World, among others. Pugwash has, moreover, for many years extended its remit to include problems of development and the environment.[96] The Pugwash Conferences on Science and World Affairs were awarded the Nobel Peace Prize in 1995.

Pugwash as an example of the service of science to foreign policy will be treated in 5.4.

4.2.6   The German-American Academic Council Foundation (GAAC)

Another institution to be mentioned here is the *German-American Academic Council Foundation (GAAC)*. Founded following a joint announcement by *Chancellor Kohl* and *President Clinton* in 1993, the foundation is dedicated to strengthening German-American co-operation in all fields of science and the humanities, particularly by bringing together and utilizing the experience, expertise, and commitment of its members. The Council provides a forum for transatlantic dialogue, conducts policy studies of mutual interest to decision makers in both countries, and encourages development of collaborative networks, especially of young scientists and scholars.[97] So far, working groups have completed studies on the elimination of excess weapons plutonium, and on German and American migration and refugee policies. Symposia were held on academic and policy issues confronting Germany and the United States, on the future of the Humanities and Social Sciences in

---

[96] *loc.cit.* (Footnote 6).
[97] From the impressum of GAAC report U.S.-GERMAN COOPERATION IN THE ELIMINATION OF EXCESS WEAPONS PLUTONIUM, National Academy Press, Washington, D.C. 1995.

Central East European countries (CEEC) with consideration of experiences from the German unification process, on New Horizons in Research and Higher Education (Trends, Constraints and Opportunities) and on Scientific Research in Universities, Academies and Extra-University Institutions of the CEEC, Germany and the United States - Approaches, Experiences, Perspectives. For German scientists GAAC offers an opportunity for taking part in interdisciplinary investigations of pressing problems. The U.S. NAS has provided such opportunities to American scholars for many years, but in Germany there is no national academy of sciences. The GAAC partly fills this gap.

4.3   The professional culture of scientists in negotiations

An education in one of the natural sciences creates a certain mentality. In 4.1 this subject has already been touched. Those who have undergone this education have learned to respect facts and the laws of nature which govern the physical world. They know that wishful thinking does not help to obtain results but that conscientious work is required. They know from experience that machines will stop running sooner or later unless they are supplied with energy and maintained carefully. They have found out that there is a difference between mere assumptions and firm knowledge, and they are familiar with the concept of probability. They can distinguish between predictable and unpredictable courses of events, and between developments that are likely to occur and those the occurrence of which just cannot be excluded. They respect proofs and are willing to change opinions that are proven wrong. They are convinced that cheating does not pay because, usually rather soon, fraud is detected by those who try to confirm the feigned result. Most scientists are familiar with international co-operation and have friends in other countries with whom they have exchanged results or cooperated in joint projects. This has made them immune against national prejudices.

What is the influence of these attributes, and of those mentioned in 4.1, on a scientist who is asked to act as negotiator? The present author had that experience when, for instance, he had the task of reassuring the U.S. State Department and the U.S. Atomic Energy Commission (AEC) about Germany's continued reliable partnership as a buyer, from the United States, of slightly enriched uranium for German nuclear power plants when

Germany, in the early 1970s, decided for the first time to buy a certain amount of this fuel from the Soviet Union where it was available at a lower price. The U.S. Atomic Energy Commission (AEC), Germany's sole supplier till then, was upset, as was the State Department. They were wondering whether, under the influence of the *Ostpolitik* of *Chancellor Willy Brandt*, the Federal Republic of Germany was slowly drifting away from the Western camp. It was possible to reassure them that this purchase was an entirely commercial, market-oriented step, and that German power plants would be happy to continue buying U.S. uranium, particularly when the price came down.

Before joining the staff of the German Embassy in Washington, D.C., the author had been a division leader in an independent research institute, and was used to make his own decisions, after consultation with his colleagues, about the experiments to be carried out, and the equipment to be made or bought. Sometimes negotiations had to be made with firms, or with the staff of the large accelerator centers about machine time at the accelerators where the equipment was to be exposed. Usually, there was collaboration with groups in other countries, and that required the negotiation of agreements about the division of the data obtained, the standards of measurement, etc. A certain amount of bargaining was necessary occasionally but, in general, agreement was easily reached. All had the same interest in a successful scientific experiment giving interesting results.

As Counselor for Scientific Affairs (also called Science Attaché) at the Embassy the author had to learn some new rules.[98] First of all, when he wrote a report about his work, about the information he had received, or the results of his negotiations, it was the Ambassador, not he, who signed the report. The Ambassador might even change some of the wording into a more diplomatic language. And the report always had to go to the Foreign Ministry in Bonn, with a copy to the Ministry of Research and Technology, and not to anybody else who might also be interested. Who else would be informed was decided in Bonn.

After a while the author got used to these and some other rules. He understood that scientists, in order to be effective negotiators and/or government advisers, must be familiar with the administrative procedures of governments, with the rules of international diplomacy, and with the political

---

[98] K. Gottstein: Forschung, Technologie und Politik. Die Aufgabenbereiche der Wissenschaftsreferenten an den Deutschen Botschaften. Physikalische Blätter 29, Heft 10, Oktober 1973, S. 435 – 444.

obstacles that may have to be overcome if satisfactory solutions, or agreements, are to be reached. On the other hand, he discovered that his previous training as a scientist was very useful in his diplomatic work. When a particular issue arises, a professional diplomat is inclined to concentrate on finding the best way to represent the position of his government on that issue. He is trained to consider his own opinion to be of lesser relevance. The average scientist, however, is trained to find out for himself about the merits of a case, and to form his own opinion. This enables him to give independent advice to his superiors. Even though his instructions will not always allow him to express his opinion freely at the negotiation table, the fact that he has formed one independently gives him greater confidence and a self-assured behaviour. His opponents will be more inclined to take him serious, and his own government may be willing to take his suggestions into account. There is often some flexibility in the position to be negotiated which an intelligent negotiator can use for the benefit of his own side. Moreover, the informality, openness and the friendly behaviour that usually characterize discussions among scientists and to which scientists, unconsciously, are used, are helpful in creating a climate in which even difficult problems can be discussed productively. Cultural differences that usually make mutual understanding difficult play a lesser role among scientists who share the common culture of science. In addition, they share the "elite culture"[99] of those who travel a lot, know several languages, and are used to the exchange of ideas and arguments. Thus, the scientific background can be of assistance to scientists for obtaining satisfactory results in diplomatic negotiations.

It should not be concealed, however, that a certain understanding broadmindedness on the side of his superiors is sometimes required to take full advantage of these assets. Because of their independence, scientists can be reluctant to follow rules which they do not understand or of which they do not approve. At least, they want to have an explanation why it might be considered reasonable to accept a particular rule. Rather than offering that explanation, or waiving the rule, it is sometimes preferred to look upon scientists as "eggheads" who lack political understanding and cannot be fully relied upon to follow instructions. Under this judgement, scientists may be qualified as government representatives when their task is well defined and lies within their own field of competence, but otherwise it would be better not

[99] W. Lang: The concept of "professional cultures" as a tool for the better understanding of international negotiations, IIASA PIN project, unpublished, page 6.

to pay too much attention to them. For example, *Professor Werner Heisenberg* was appointed to sign, on behalf of the government of the Federal Republic of Germany (FRG), the international convention by which the European Organisation for Nuclear Research (CERN) was created. But when, as one of the 18 signers of the "Göttingen Declaration" of 1957, he advised the same government to refrain from trying to acquire nuclear weapons he was severely criticized for being naive.

Fortunately, this type of skepticism regarding the usefulness of scientists in political negotiations with a scientific background is not universal. Particularly in the United States it is quite common for scientists to change back and forth between teaching and research, government service, and work for industry. This is looked upon with approval by the scientific community, and the government as well. In Germany this kind of flexibility is still rare, but it does occur.

5 Science in the Service of Foreign Policy

5.1 Example 1: The International Institute for Applied Systems Analysis (IIASA)

An example of how not only individual scientists but science as such can serve the purposes of foreign policy is the *International Institute for Applied Systems Analysis (IIASA)*. It was founded in 1972 during the Cold War, at the initiative of the USA and the USSR, in order to reduce the tensions between the two opposing blocs by scientific co-operation on problems of general and urgent interest. A description of its programme was given in section 2.9.4 of chapter 20.

5.2 *Example 2:* The Scientific Forum of the CSCE

Another example of science serving simultaneously the purposes of foreign policy and its own goals of international co-operation in applying scientific methods to the solution of universal human problems is the *Scientific Forum of the Conference on Security and Co-operation in Europe (CSCE)*[100]. It was the then Foreign Minister of the FRG, *Walter Scheel*, who suggested in Helsinki in 1973 to organize a Scientific Forum as a means of *détente* during the Cold War. This suggestion found its way into the Final Act of the CSCE, adopted in 1975. The government of the FRG offered to organize a meeting of experts of the 35 CSCE countries in Bonn. This was accepted. The meeting lasted six weeks (June 20 to July 28, 1978) and, after many controversial discussions, produced an agenda for the Scientific Forum and decided about its date and place. The Forum itself was held in Hamburg in February/March 1980 and lasted two weeks. 35 delegations attended, mostly consisting of prominent scientists, accompanied by some science administrators and diplomats. On the agenda were problems in areas of science in which international co-operation was desirable: research in

- alternative energy sources,
- food production,
- cardiovascular, tumour and virus diseases, taking into consideration the influence of the changing environment on human health, and
- social, socio-economic and cultural phenomena, especially the problems of human environment and urban development.

It must be remembered that both the "meeting of experts" and the "Scientific Forum" took place at a time when the Cold War had reached renewed intensity. The arms race continued, Afghanistan had just been invaded, dissidents were persecuted in the Soviet Union, scientists in the West resented the travel limitations which the Soviet Union imposed on its own scientists as well as on foreign visitors. Both the meeting of experts and the Forum were close to breaking up on several occasions when Western participants accused the Soviet Union of violating human rights, using the banning of Sakharov to the city of Gorki as an example, and Soviet participants responded by pointing to the allegedly continuing discrimination

---

[100] K. Gottstein: Das "Wissenschaftliche Forum" der Konferenz über Sicherheit und Zusammenarbeit in Europa (KSZE), Wirtschaft und Wissenschaft 27, Heft 2/1979, p.18 ff.

K. Gottstein: Das Wissenschaftliche Forum der KSZE. Verlauf, Ergebnisse, Ausblick, Physikalische Blätter 37, Nr. 2, 1981, S. 48 – 49.

of blacks in the United States. All along it was doubtful whether it would be possible to reach consensus on a final report of the Forum. Surprisingly, consensus was reached after all. A list of concrete proposals for co-operation was produced. Western scientists, after having expressed their dismay at Soviet violation of human rights, and Eastern scientists, after having dutifully repudiated these reproaches with accusations of their own, were at last united in their desire for an improvement in international co-operation in science. It was said that the congenial atmosphere, created for the expert meeting as well as for the Forum, with the receptions and excursions customary for "normal" science congresses, had contributed to this success. As chairman of the West German delegation at the Bonn expert meeting and as Executive Secretary of the Hamburg Forum the present author had tried from the beginning to let the national delegations feel that they were all equally welcome, and that the Forum was n o t just a pretext for political purposes but a serious scientific enterprise, the success of which was earnestly aspired. The fact that he was personally acquainted with several heads of delegations, both from the East and the West, was certainly helpful.

5.3 *Example 3:* **The United Nations Conference on Science and Technology for Development**

In between the 1978 Bonn expert meeting and the 1980 Hamburg Scientific Forum of the CSCE the author had another opportunity to notice the particular value of personal relations between scientists in diplomatic negotiations. At the *United Nations Conference on Science and Technology for Development (UNCSTD)*[101] he was co-ordinator of the scientific contributions of the FRG to that conference (Vienna, 1979) and in this capacity a member of the official West German delegation which was led by the Federal Minister for Research and Technology, *Volker Hauff.* His opposite number in the delegation of the German Democratic Republic (GDR) was *Professor Claus Grote*, like himself a high energy nuclear physicist. Indeed, about 20 years earlier the author had helped clearing the ground for the admission of the GDR high energy nuclear physics group, to which Grote then belonged as a young research physicist, to experiments at

---

[101] K. Gottstein: Science and Technology for the Third World: The United Nations Conference on Science and Technology for Development. Economics <u>21</u>, 136 - 151, 1980.

the CERN accelerator.[102] (The GDR was not a member of CERN.) It was not surprising that, when they met again in Vienna, the author had very easy relations with Professor Grote even though their two delegations belonged to the two different sides in the Cold War.

5.4 *Example 4:* The Pugwash Conferences on Science and World Affairs

The ***Pugwash Conferences on Science and World Affairs*** have already been mentioned in 4.2.5 as an example of how a group of individual scientists from different nations advised, and continues to advise, governments about existing dangers, and how to avoid them. But the Pugwash Conferences can also be used as an example of the service to foreign policy that scientists with their special qualifications can render. Much of the effectiveness of the Pugwash Conferences during the Cold war, particularly in crisis situations, was due to the openness that is possible among colleagues who have known each other for a long time and who, because of their training in science, think along similar lines. The meals and the social events at these conferences provide opportunities for informal conversations in which questions can be asked and answers given which it would hardly be possible to ask and give in official negotiations, and which government representatives, in particular, would be unable to express out of fear that such statements could be interpreted as an indication of existing intentions, as a sign of weakness, or that they could have some other undesirable consequences. Among scientific colleagues, however, it was much easier to ask questions such as "How would your side react if our side would do or propose XY?" One knew that the colleagues had good connections to their government and that everything of interest said would be reported. If the government thought that the report was useful it could take it into account. If it thought that the suggestions made were impractical it could easily ignore them without anybody losing face. The present author happens to have quite a number of letters from the Foreign Ministry in Bonn saying that the reports from Pugwash Conferences were very interesting and helpful. Before going to a Pugwash meeting during the Cold War he used to visit the Foreign Ministry and the Ministry of

---

[102] Thomas Stange, Die Genese des Instituts für Hochenergiephysik der Deutschen Akademie der Wissenschaften zu Berlin (1940-1970). DESY-Thesis 1998-019, Hamburg 1998. ISSN 1435-8085.

Defence, to familiarize himself with their thinking, their assumptions and perceptions in order to be able to attempt verification or falsification, the creation of mutual understanding, and the removal of misunderstandings.

5.5 The International Conferences of Research Unit Gottstein in the Max Planck Society

The conferences organized by Research Unit Gottstein (on problems of science and society) in the Max-Planck-Gesellschaft served the same purpose. The *conference on the Strategic Defense Initiative (SDI)*[103] of President Reagan came to the conclusion that it depended on the assumptions regarding the intentions and perceptions of the Soviet Union whether SDI was good or bad for strategic stability. Our series of international *"Perception Conferences"*[104] clarified the assumptions, in each of the European countries, about the intentions and perceptions of the major powers as well as of the immediate neighbours of each country. These assumptions and perceptions influence, of course, opinions and policies in these countries. That is why it is so important to know about them.

6   Conclusions

Winfried Lang[105] concluded that "the role of culture in international negotiations is not restricted to the impact of national cultures and that it even may be assumed that national and professional cultures compete with each other, that the growing specialization enhances the importance of

---

[103] K. Gottstein (editor): SDI and Stability. The Role of Assumptions and Perceptions. Nomos Verlagsgesellschaft, Baden-Baden 1988, 396 pages.

[104] K. Gottstein (editor): Western Perceptions of Soviet Goals: Is Trust Possible? Campus Verlag/Westview Press, Frankfurt/Boulder 1989, 455 pages.

K. Gottstein (editor): Mutual Perceptions of Long-Range Goals. Can the United States and the Soviet Union Cooperate Permanently? Campus Verlag/Westview Press, Frankfurt/Boulder 1991, 404 pages.

K. Gottstein (editor): Integrated Europe? Eastern and Western Perceptions of the Future. Campus Verlag/Westview Press, Frankfurt/Boulder 1992, 295 pages.

K. Gottstein (editor): Tomorrow's Europe. The Views of Those Concerned. Campus Verlag, Frankfurt 1995, 845 pages.

[105] W. Lang, loc.cit. (Footnote 11), pages 7/8.

professional cultures for international relations in general and that a specific negotiation culture was about to emerge which might benefit from various professional cultures cutting across traditional lines of thinking and behaviour anchored in the national backgrounds of various negotiators." As an example of a highly professional culture, Lang mentions the so-called "lawyer-diplomat". It seems that one might also find examples of "scientist-diplomats". Admittedly, they are very much rarer than "lawyer-diplomats".

Guy Olivier Faure states[106]: "In an international negotiation, where national culture is the discriminating variable, a common professional culture can be viewed as a first step already made in the learning process that will enable the parties to reduce their divergence. In this respect, professional culture ensures a certain level of functionality with regard to this learning process, an essential part in the set up of the necessary chemistry for the negotiation progress." Again, this is certainly true for the scientific culture. It is not so much a particular choice of words, a particular behaviour, or even similar skills or philosophical beliefs that constitute the scientific culture, although all this may contribute in special cases. It is rather a respect for facts, the urge to find out about the driving forces behind the phenomenological world, the skepticism about unproven assertions, the contempt for lies, the desire to make progress and see results which are elements of the scientific culture and which sometimes enable scientists to contribute usefully to negotiations and, in special cases, even to break deadlocks.

Acknowledgements

I am much indebted to my old friend and colleague Professor Victor Kremenyuk of the IIASA PIN project who suggested that I should write this paper, and to Professors Rudolf Avenhaus and Winfried Lang for some useful suggestions.

---

[106] G.O. Faure: Professional Cultures. Concepts and Problematiques. IIASA PIN project, September 1996, page 11, unpublished.

22 The New World Order and the Role of Science

Final remarks

The preceding chapters of this book could only give an incomplete picture:

- of the nature of the risks which humanity is already running, and which it will be facing increasingly in the next century,
- of the disasters and catastrophes that might happen unless these risks are reduced as far as possible by adequate measures,
- and of the role which the institutions of science could play as sources of advice to policy-makers in this preventive risk reduction.

The risks to be taken into account in this context are:

- ♦ threats to health and human rights caused by:
- ⇒ war
- ⇒ starvation
- ⇒ disease
- ⇒ persecution
- ⇒ discrimination and social injustice
- ⇒ illiteracy in the presence of high technology
- ⇒ the storage and dismantling of nuclear and chemical weapons, the illicit production of biological weapons, nuclear proliferation, the malfunction of nuclear power
- ⇒ mental disturbances and instabilities created by fundamentalism and national chauvinism
- ⇒ mass unemployment
- ⇒ the vulnerability of our modern scientific-technological society to ignorance and unethical behaviour,

- ♦ threats to the natural environment caused by:
- ⇒ the general pollution of the ozone layer, the atmosphere, of lakes, rivers, and oceans, and of the soil

> ⇒ climatic changes due to the dying of forests, the destruction of rain forests, and the spreading of deserts
> ⇒ the extinction of plant and animal species,

♦ threats to traditional cultural identities caused by
> ⇒ the worldwide spread of Western civilization by means of the world market
> ⇒ tourism
> ⇒ migration of refugees and "ethnic cleansing"
> ⇒ television.

These threats are often due to thoughtlessness, inertia, or even criminal actions in violation of national and international law as well as of human rights. With regard to cultural identities, efforts are under way in North-South relations and in development co-operation to take them into account and preserve them as far as possible.[107]

Efforts to deal effectively with any of the threats mentioned above are meeting with powerful obstacles. Among them are:

- overpopulation in many areas of the world,
- limited resources,
- rising wants, demands and claims among the poorer parts of the world population because tourism and television enable them to compare their own standard of living with that of the affluent members of world society,
- nationalism combined with ideological prejudices, enemy images, and archaic instincts of aggressive behaviour.

In the long run it will not be possible to cope with the aforementioned threats, minimizing the risk of the occurrence of catastrophes, without making compromises, such as:

---

[107] See, e.g., Klaus Gottstein (Ed.), Cultural Development, Science and Technology in Sub-Saharan Africa, Nomos Verlagsgesellschaft, Baden-Baden 1986; Klaus Gottstein (Ed.), Islamic Cultural Identity and Scientific-Technological Development, Nomos Verlagsgesellschaft, Baden-Baden 1986.

- the compromise between efforts to reach a high standard of living, and efforts to protect nature and the global environment,
- the compromise between the effort to win a struggle for power and the effort to avoid (civil) war.

In the first case, this may require the readiness to pay a price for the preservation of some remnants of original nature and for environmental protection. The establishment of large "Global Parks", analogous to the National Park system in the United States, would be one way to preserve living space for endangered species of animals, and to protect rain forests. Human populations living there would have to be compensated for their sacrifice of not being allowed to develop these areas from which they may have to be removed.

In the second case, World Courts and World Police Forces will be required, or will have to be made effective, so that political conflicts can be settled by arbitration and jurisdiction, if bilateral negotiations fail, and the execution of judgements can be enforced.

If measures of this type for the prevention of catastrophes are to have some chances of success, it seems obvious that, in the long run, some prerequisites must be fulfilled:

- the increase in world population must be stopped before an unbearable situation arises,
- international collaboration in the analysis of problems and in the design of, and in the firm agreement on preventive or curative measures must be intensified and institutionalized,
- public education must be improved so that the majority of the population realizes what the problems are, and supports politicians who work for long-range solutions,
- regional units transcending national borders must be created to overcome nationalism, and to give equal chances to inhabitants of economically "backward" sub-regions or countries so that they become citizens of more competitive units within which they have a chance to move freely and to work without the restrictions necessarily existent in small, politically and economically isolated countries.

This is a large and very ambitious programme which, of course, will need many amendments in the course of time. It can be realized only step by step over many years. But time is running out in many respects, and our efforts must increase.

Many questions are still open and need further research. Among them is the fundamental question to what extent economic considerations can be allowed to govern, without social control, the policies of decision-makers. In the last analysis, what do human beings need to feel content? We all know that we tend to compare our lot with that of the people around us, and that we feel unhappy if theirs seems to be better than ours, even if ours is tolerable, because we believe that they look down on us. For this reason black people in South Africa, during *apartheid*, were not reconciled with their situation by being told that they had a higher standard of living than their brothers and sisters in the independent nations around South Africa. This is not what counted. South African blacks wanted to have the same rights, and be entitled to the same respect, as the white citizens in their midst. The absolute standard of living was of secondary importance, the relative standard counted as a measure of self-esteem.

What does this mean for the world society of the future when, due to modern means of communication and transportation, the affluent minority of the world population will live "in the midst" of the large majority of poor people? Psychology has a fertile field of research here, and politicians as well as economists will be well advised to study the results very carefully.

Psychology as well as the disciplines of anthropology and history are also called upon when it comes to the reasons why neighbouring groups of people sooner or later start fighting each other unless certain conditions are fulfilled. This is discussed in chapter 8. Since lack of mutual trust, often based on inherited enemy images is at the roots of many conflicts it is shown there how trust among former opponents can be created by working on joint projects when both sides are threatened by the same danger. In previous history this danger was often a common enemy. Today we must learn that all of humankind is threatened by environmental and other global problems which can only be mastered by joint efforts, forgetting the enmities of earlier generations. These joint efforts must include interdisciplinary studies, carried out in international co-operation of scientific institutions, directed at the development of options for political action. Decision-makers and the public must be informed as to what the available options are. The description of each option should be accompanied by an estimate of the chances, risks and costs connected with it. For each measure recommended the potential technical, financial, social, psychological, political consequences should be clearly stated so that the public and the decision-makers know what to expect when certain political actions are taken. Many catastrophes could be avoided if risks were properly assessed and taken serious and necessary precautions were not avoided for financial reasons, just hoping for the best. The obstacles must be studied which often prevent the implementation of recommendations given by the technical experts with respect to each of the global problems. This means that legal experts, historians, psychologists, social and political scientists must be involved.

In many parts of the world an important obstacle to orderly procedures is the lack of functioning democratic institutions. Can a free democracy work satisfactorily only when economic conditions are good and when people have reached a certain level of education, enabling them to distinguish between different political platforms? Aren't there counterexamples in history, both of poor countries with democratic government systems and of well-to-do countries with authoritarian rule? These are questions that need further clarification by political scientists, economists and historians.

Considering that a large proportion of the earth's population will remain poor for a long time, and that responsible democratic governments probably cannot survive among starving people, how can the world economy be induced to produce the goods required to cover the basic needs of the world population, and to deliver them to the places where the need is? What kind of subsidies, from local, regional or international sources, would have to be raised for this purpose? Again, interdisciplinary studies by economists and political scientists with knowledge of local conditions could produce answers to these questions.

Many questions remain to be answered in the fields of arms control, conflict resolution, environmental protection, preservation of nature, and sustainable development. All these questions have scientific components. Without clear scientific advice, politicians and the public will be at a loss in many instances as to what the best decisions are under the existent circumstances. As shown and discussed in several places in this book, they will have to be told what the options are, and what the likely consequences are going to be for each option: the chances, the risks, the costs.

Already in the preface to this book, and in many of its chapters, we have called attention to the fact that here lies a task for the international community of institutions of science. They have already begun to shoulder this task in various ways, as was shown, e.g., in chapter 20. But much more could be done. Standing international and interdisciplinary working groups could be set up in which leading experts constantly monitor developments causing concern because of their potentially dangerous trends and consequences. These working groups should elaborate options for action, giving estimates of chances, risks, and costs for each action, as explained in this book.

There is hope that the international scientific community will finally be able to organize an advisory mechanism of this kind. If this hope materializies, humankind will have a chance to prevent some of the worst catastrophes and conflicts which otherwise, unfortunately, are likely to occur.

23 Summaries

a. Questions needing further research

Condensed list of problems to be analyzed

(see p. 11/12)

- Risk assessment, risk-benefit analysis and discussion of possibilities for international sharing of great risks (oil supertankers, nuclear power stations, climatic changes by the destruction of rain forests, etc.) In many cases, risks of this type could be strongly reduced by expensive precautionary measures. The obligatory introduction of such measures would require international agreements on rules and regulations for the reduction of risks, and for the distribution of the financial burden involved,
- Determination of the size and quality of reservations and natural habitats necessary for the preservation of animals and plants threatened by extinction,
- Rough determination of the minimum volume of biomass (trees and plants) required for the preservation of a balanced relation of the combinations of oxygen, carbon and nitrogen in the atmosphere and for the prevention of undesirable changes of climate,
- Drafting of a list of geographical regions which, according to the criteria mentioned above, would have to be exempted from industrialization and dense population,
- Juridical and economic investigations about possibilities for compensating inhabitants of regions which, because of the protective measures indicated above would be denied development and industrialization, and might even have to be resettled,
- Investigations by experts of international law on possibilities for the protection of such reservations from poachers and other intruders (Green Helmets of the United Nations?),
- Questions of the conversion of arms factories to civilian production,
- Economic problems of East-West co-operation,
- Special topics of scientific-technological co-operation between industrialized and developing countries.

Managing risk

(see page 9)

International co-operation, if it is to be successful, requires psychologically trained understanding of cultural differences and of differences in perception. This is particularly true for the estimation of risks from so-called human failures in the operation of large-scale technologies such as nuclear power stations, oil supertankers, arsenals of ABC weapons. In cases of this type it is not sufficient to set arbitrary safety factors in order to minimize the risk of accidents. It will rather be necessary to investigate all possibilities for errors in the behaviour of all members of all potential operation, maintenance and repair teams, taking into account their cultural backgrounds and psychological attitudes. Even after appropriate measures were taken it will be necessary to include a still thinkable "worst possible out-turn"[108] into the analysis. "Managing risk" is a new science which will be of increasing importance in years to come.

Economic problems of East-West co-operation

(see pages 8, 12, 149)

In economic policy international co-operation is also necessary because laws in the field of environmental protection must be valid worldwide if they are to be effective. A new danger to stability has arisen: exaggerated "friend images". It derives from the over-optimism of some interactionists who expect co-operation to lead to immediate success in the economic field. Expectations   that are unrealistic may lead to disillusionment, public frustration, disappointment and, as a consequence, to political instability.

North-South co-operation

(see pages 2, 12, 222)

The economic gap.
Special topics of scientific-technological co-operation between industrialized and developing countries.

---

[108] Benton, Peter, Riding the Whirlwind, Oxford 1990, page 158.

The role of cultural identities.

Perceptions of economic threat

(see page 157)

Are there new threat perceptions in the economic area? By what kind of co-operation could they be possibly overcome?

What do humans need to feel content?

(see page 224)

Many questions are still open and need further research. Among them is the fundamental question to what extent economic considerations can be allowed to govern, without social control, the policies of decision-makers. In the last analysis, what do human beings need to feel content? We all know that we tend to compare our lot with that of the people around us, and that we feel unhappy if theirs seems to be better than ours, even if ours is tolerable, because we believe that they look down on us. For this reason black people in South Africa, during *apartheid*, were not reconciled with their situation by being told that they had a higher standard of living than their brothers and sisters in the independent nations around South Africa. This is not what counted. South African blacks wanted to have the same rights, and be entitled to the same respect, as the white citizens in their midst. The absolute standard of living was of secondary importance, the relative standard counted as a measure of self-esteem.

What does this mean for the world society of the future when, due to modern means of communication and transportation, the affluent minority of the world population will live "in the midst" of the large majority of poor people? Psychology has a fertile field of research here, and politicians as well as economists will be well advised to study the results very carefully.

Conditions for a working democracy

(see page 225)

In many parts of the world an important obstacle to orderly procedures is the lack of functioning democratic institutions. Can a free democracy work satisfactorily only when economic conditions are good and when people have reached a certain level of education, enabling them to distinguish between different political platforms? Aren't there counterexamples in history, both of poor countries with democratic government systems and of well-to-do countries with authoritarian rule? These are questions that need further clarification by political scientists, economists and historians.

b. On the psychology of friendly and enmical groups

Co-operation and confrontation

(see pages 2, 86, 88)

Human beings are willing to co-operate with those whom they recognize as members of their own tribe or group. With them they are prepared to work together for common goals and joint enterprises. But they are equally inclined to fight bitterly against those whom they consider to be members of an opposing group, competitors for food, land, or work, or just suspicious "aliens" who do not share one's own beliefs, customs, habits, or values. They are declared to be enemies. Such "enemy images" are often upheld for centuries. Interestingly enough, however, peoples who have fought each other for centuries, like France and Germany, can become friends when a common enemy appears who seems to threaten all of them. Then they can be willing to join forces in order to meet the new challenge.

Internal problems of humanity as the common enemy

(see pages 3, 28)

The inhabitants of this planet will have to learn that they all belong to *o n e* group which has to keep peace and co-operate because it is threatened by a common enemy. This time the enemy is not coming from outside but from inside. It is the misuse, or the thoughtless use, of science and technology, irrespective of the undesirable by- and after-effects of often well-intended applications of science and technology. It is also the submittance to archaic instincts against people who do not seem to share our values although these people are our co-passengers in a boat struggling to stay afloat in stormy weather and needing all hands for bailing out water.

Conditions for keeping peace among feuding ethnic groups

(see page 26)

1. The use of arms must remain the monopoly of an authority at a level higher than that at which the potential enemies are located. Otherwise the war, when started, will probably be fought to the bitter end. In the case of Yugoslavia, e.g., this would mean that the war-fighting parties would have to be disarmed and that the embattled areas would have to be policed by an international (United Nations or European) force until a reliable and enforceable peace settlement had been found. This international police force would have to be ready to use its arms for the protection of peace.
2. Parties to the conflict must be made to agree to settling their differences by bringing their cases to an International Court of Justice, and by recognizing as binding the judgements passed by this Court. Its decisions must be enforceable.
3. All possible steps must be taken to reduce existing enemy images which are the source of hatred, misunderstanding and fear.

Elimination of enemy images by explaining their origin

(see pages 27, 63)

Psychologists, historians and political and social scientists can help with the elimination of enemy images. They can explain the origin of threat perceptions and the way in which fears and misunderstandings reinforce each other and lead to preventive or retaliatory actions which are interpreted as aggressive by the other side and thus seem to confirm the original enemy image. They can also explain, to the general public, the role of enemy images for rallying the public behind a demagogic leader. Sometimes enemy images derive from a feeling of superiority which is cultivated by minority populations with respect to the majority around them and which serves to prevent intermingling and thus to preserve their cultural identity. This can be explained, made conscious and thereby deprived of its potential explosiveness.

Elimination of enemy images by joint efforts on global problems

(see pages 28, 224/225)

In addition to explaining, and putting into perspective, historical, psychological, geographical and other scientific facts underlying enemy images, scientists and scholars must call attention to the global and universal problems of environmental pollution, hunger, disease, drugs, population growth, energy shortage and related problems which really need every conceivable effort. The group feeling of solidarity created by enemy images should rather be directed towards these threatening global problems. Scientists and scholars have an important role here, not only in describing these problems but in designing strategies for dealing with them. These strategies should be based on interdisciplinary studies by local, national and international experts. They should assess by-effects and after-effects of measures considered for implementation, as well as the risks and benefits involved. This should result in options for action. Co-operation of different factions of the population in the discussion of these studies and in planning the implementation of measures could create feelings of partnership and assist in the elimination of enemy images.

Discussion of options for action

(pages 44/45)

The options for action to be worked out in scientific detail must be compatible with each other. It is well known that they must concern very diverse fields, such as:

- the control of the armaments of groupings at enmity with each other and the design of CBMs fit for overcoming the enmity
- the design of joint measures, e.g. for agricultural and industrial development, with an optimal protection of the environment
- joint measures for an improvement of medical services, birth control and nutrition
- joint measures for education and professional training, particularly of women, with special attention to the need for the maintenance and repair of urgently needed medical and technical equipment, devices and machinery.

But the options must find attention and be acted upon. This will only happen in an appropriate way if joint discussions of politicians, economists, educators, philosophers, psychologists, theologians are institutionalized.

The network of fear-creating problems

(see page 55)

If dangerous threats to peace are to be averted, then the study of methods of arms control though important, is not enough. The whole network of fear-creating problems will have to be tackled in a systematic and coordinated manner.

Solidarity cannot become effective without knowledge of the facts that have to be taken into account in practising it. In many threats to peace these facts are very complicated, pertain to different disciplines of knowledge, and are under the jurisdiction of several nations. They must, therefore, be tackled in interdisciplinary and international co-operation.

New forms of living together

(see page 29)

In the long run, modern technology will lead to close contact between all races, religions and cultures. New forms of living together will have to be developed. We should try to start in those places where different nationalities have lived together for a long time. There we should eliminate tensions as far as possible. But we should not divide people from each other artificially by removing them from their native lands because this leads to uprooting, alienation and to a destruction of cultural values. Under no circumstances should existing social and political structures be dissolved before general agreement has been reached on the new structures which are to replace them, and before these structures work.

Aggressive behaviour as the result of fear

(see pages 54, 79)

Threatening behaviour of persons can often be explained as the result of fear. This is also true for nations. Self-hate, due to one's own imperfections, is projected onto the others. The fight for justice presents itself with the pathos of hatred. This hatred justifies fear, and it justifies the hatred of the opponent.[109] Since fear tends to create an aggressive state of mind, in order to compensate for the insecurity connected with it, we must realize that we cannot deal in an isolated fashion with nationalism or armaments (which are just symptoms of fear) if we want to treat threats to peace in a general way. We must look at all the major sources of fear which beset people - politicians, governments and the public - at this time in history. Fear is also produced by distressing situations such as overpopulation, environmental disasters, social inequalities, the gap between the standards of living of the North and the South. What is needed, is an open discussion of threat perceptions.

Conditions for the appearance and 'cure' of enemy images

(see page 61)

Theoretically, the best solution for conflicts threatening to get out of control and turn violent would be, of course, to remove the enemy images on which a beginning feud is based. Can this be done? These enemy images might have their origin in a variety of genuine or merely perceived conflicts of interest, or in racial prejudices, in traditional antagonisms between neighbouring, competing tribes or groups, in imagined irreconcilable religious differences, etc. What is known about enemy images and the methods that could prevent their appearance and growth? Is it possible to 'cure' enemy images, and if so, what are the prerequisites for the efficacy of these methods? What can be said regarding the available methods of crisis management and conflict resolution when violent eruptions are to be avoided between groups confronting each other, and when peaceful settlements are to be reached? Under which conditions do social and political changes of the type observable in former Yugoslavia and in the former Soviet Union lead to aggression against minorities, to racism, ethnic cleansing, extreme nationalism, civil war? What role does the weakening, or the fading away, of central authority have in this respect? How do the enemy images originate which underly armed conflicts between populations which, more or less, have lived together peacefully as long as a strong central authority existed? Which psychological tools are available for replacing by cooperative behaviour those enemy images and the racism, chauvinism, xenophobia connected with them?

It would be highly worthwhile to take stock of the available answers to these questions.

---

[109]C.F. von Weizsäcker, Der Garten des Menschlichen. Beiträge zur geschichtlichen Anthropologie, page 475. Carl Hanser Verlag München 1977.

Condensed conclusions

(see pages 66-68)

- Independent groups of people develop independent ideas which, in time, lead to conflicts of interest.
- If there is no superior authority to which quarrelling parties can appeal and whose judgement is accepted and enforceable, people will feel entitled to use force. They will fight unless there is mutual deterrence of the kind that fighting clearly would mean self-annihilation of both sides. This is true for individuals as well as for ethnic groups or nations. It is the cause of wars.
- Former enemies can be become friends and allies when they perceive a common enemy threatening both of them.
- The fading of enemy images can be greatly assisted by studying the chosen traumas and glories (Volkan) of the opposing parties and by making each side understand them. Both sides must learn to see the situation through the eyes of the other side.
- People of different races and different languages can live together peacefully in one state if they do not feel oppressed and if they respect the institutions of their nation or state. Brazil and Switzerland are examples.
- Peoples which have lived together peacefully under common rule - as the peoples of the Ottoman, Habsburg and Romanov empires - can become enemies when that central rule crumbles (the first two theses then apply).
- If there is to be a dissolution of joint institutions, peace can be preserved if a new, stable power structure is set up before the old one is abolished, and if the transition is made under pre-arranged agreements. Czechoslovakia's dissolution is an example.
- In order to prevent further wars the setting up of strong mediation, peace keeping and peace enforcement authorities should have high priority. Whether these can be developed by the United Nations independently, or with the assistance of NATO, OSCE, European Union or new organizations, is to be studied.
- In the past, joint tasks like hunting a mammoth, defending a castle or building and defending an empire, have motivated people of different origin to join forces and forget their differences. Today, there are global problems threatening the health and life of all of humankind. People should be made aware of the fact that these problems can only be solved

by world-wide co-operation and that all the local and national causes for war are minor issues which should be set aside in the interest of joint action for the protection of life on our planet.

- Peace and stability in a multi-ethnic society - and all our societies will become multi-ethnic in the 21st century due to world-wide migration - can only be maintained if education to tolerance and ethical behaviour starts at an early age and continues throughout life, by the school system, the churches, the media, the law system, the trade unions, etc. Without a general consensus on some general and basic rules and principles of behaviour no peaceful and stable society is possible. People must be constantly reminded of these rules and principles. Prominently among them rank the following:

  - honesty
  - readiness to help others who are in distress
  - no private use of force; law enforcement only by the central authority
  - even a noble purpose does not justify unethical means.

Taboos as a source of aggression

(see pages 67, 86)

Psychoanalysts and students of human behaviour have shown that the instincts of aggression and destruction, just as well as the instincts of love and co-operation, are deeply seated in human nature. They are the result of a struggle for survival in a hostile environment which lasted thousands of generations. In its course man learned how advantageous it was - in hunting, in agriculture, in war - to co-operate with other members of the tribe. A division of labour, a co-ordination of efforts led to much better results. Now jobs could be done, tasks could be tackled which were completely out of reach for an individual or a small group. The larger a co-operating group was, the greater was its power, given equal levels of cultural development. But in order to keep peace within the group, taboos had to be set up and aggressive instincts suppressed. As *Sigmund Freud* tells us, this suppression leads to recurring eruptions of aggression against those not belonging to the group, such as minorities and neighbouring groups.

It is obvious that we must try to re-orient our instincts before it is too late.

Common interests

(see pages 87/88)

History shows that wars between former enemies can be avoided if, and only if, theses former enemies develop a feeling of belonging to a single group which has common interests.

Conditions for keeping peace among competing groups or nations

(see pages 98/99)

Every policy, military as well as economic, of both sides, should show the entirely defensive position each side holds.

Each action to be taken should always be seen through the eyes of the other side. Could it be misconstrued by the other side as preparation for a dangerous offensive action? If that were conceivable, every effort should be made to remove such misconceptions.

Each side should recognize, and respect, the differences in the historical and social backgrounds which distinguish them from each other.

The importance of mutual perceptions

(see pages 100, 122)

How does each side perceive itself? How does each side believe it is perceived by the other side? In particular: What do we expect from each other? What could be done, in the long run, to avoid crises and improve co-operation under existing perceptions? In what respect are the perceptions changing? If it is co-operation that is required for solving global problems, what type of changes are required in the mutual perceptions of "Grand Strategies"? Which measures would have to be taken by either side if existing perceptions were to be changed intentionally in a direction representing greater trust?

We must behave as if mutual trust already existed, and set up all kinds of joint projects which need mutual trust for their functioning.

The arguments of the other side have to be taken serious

(see page 145)

The hard-liners on both sides can, if at all, only be converted by an in-depth approach to the arguments supporting their adversary perceptions, not by mere recommendations for actions following from interactionism. So each side should study the arguments of the essentialists of the *other* side, and investigate the best way to deal with these arguments.

c. General dangers

Human failure

(see page 1)

After the occurrence of great technical catastrophes - the Chernobyl reactor core melt-down, the Seveso poisoning, the Alaska coast pollution by a stranded oil tanker, the disastrous derailment of a high-speed railway train near Eschede in Lower Saxony are examples - investigations are usually started into the potential causes. Why did it happen, and how? Could it have been prevented? Frequently, human failure is identified as one of the causes. Somebody has made a mistake, somebody did not follow the rules, somebody did not inspect carefully enough the technical readiness of the equipment used.

Life-style of the North

(see page 2)

The life-style of the North, with its high consumption of non-renewable resources and its wasteful use of energy, is already causing serious damage to the natural environment. It cannot be a model for the growing world population because the balanced system of life on this planet would collapse if additional billions of people would adopt the life-style of North America or even Europe. Climate changes, destruction of the ozone layer, extinction of species of animals and plants, floods, droughts and hunger are forebodings of future catastrophes that are likely to happen unless the course of events is altered drastically.

Dangerous world situation

(see page 12)

Population growth, weapons of mass destruction, environmental destruction, and social tensions between the rich and the poor in developing countries, and between the affluent countries of the North and the majority of humankind living in the hungry Third World, have contributed to a situation in which the total system representing the human race is greatly endangered. Modern methods of transportation and communication see to it that catastrophes in one part of the world do no longer, as in earlier times, leave the other parts undisturbed. Today, television reports, economic repercussions, and masses of arriving refugees make them felt in other regions as well, in a destabilizing way.

Common enemies of humankind

(see page 19)

Hunger in the Third World, deforestation, destruction of the ozone layer, risks from inadequately designed and maintained nuclear reactors, the rapid progress in the extinction of species, illiteracy in the presence of high technology, drugs and diseases are only a random selection of those problems that may be considered common enemies of humankind.

To the creation of strong institutions of conflict resolution there is, in the long run, no alternative but war

(see pages 19/20)

In the presence of the conflicting interests and of the problems of our time, there is no alternative but war and destruction to the creation of effective international institutions entitled to make enforceable decisions.

Global and universal problems

(see page 28)

The global and universal problems of environmental pollution, hunger, disease, drugs, population growth, energy shortage and related problems really need every conceivable effort.

UNCED list of urgent problems

(see pages 35, 42)

UNCED produced a list of urgent problems regarding the protection of the atmosphere, of freshwater resources, of biological diversity, as well as regarding consumption patterns, demography, human settlement, combating deforestation, the sound management of biotechnology, of toxic chemicals and of hazardous wastes, to mention only a few problems. Many of these problems are interrelated.

Long-range dangers

(see page 37)

The long-range dangers threatening mankind by the related phenomena of overpopulation, hunger, migration, social instabilities, nationalism and fundamentalism, underdevelopment, environmental degradation and pollution, armaments, civil wars, technical failures, energy shortages, unemployment, overconsumption etc. are well known. Technical solutions are available, in theory, for most of them. But often these solutions are not compatible which each other, and therefore they are not applied.

"Seven Cardinal Threats"

(see pages 41/42)

(1) the pollution of the atmosphere, (2) the pollution of lakes, rivers and oceans, (3) the pollution of the soil, particularly by agriculture, the spreading of deserts and the dying of forests, (4) overproduction and unemployment, (5) problems of the

Third World and of the contrast between poverty and affluence, (6) "spiritual pollution" (nationalism, fundamentalism), (7) the population explosion.

Threats to all of humanity

(see pages 65, 69/70)

- The population explosion, hunger, deforestation, the extinction of species, illiteracy in the presence of high technology, drugs and diseases;
- Mental disturbances and instabilities created by fundamentalism and national chauvinism;
- Social consequences of mass unemployment and of the increasing gap between poverty and wealth, both within individual nations and between different nations;
- The vulnerability of our modern scientific-technological society to ignorance and unethical behaviour;
- General pollution of the ozone layer, the atmosphere, of lakes, rivers and oceans, and of the soil; the dying of forests and the spreading of deserts;
- Risks connected with the storage and dismantling of nuclear and chemical weapons, with the illicit production of biological weapons, with nuclear proliferation, with the malfunction of nuclear power stations and with wreckages of supertankers carrying oil.

Today's global problems

(see pages 167, 207)

- pollution of soil, water and air
- destruction of the ozone layer
- heating of the atmosphere
- desertification
- disappearance of animal and plant species in alarming numbers
- the human population explosion
- food and energy shortages
- migration of millions of people

- nationalism, racism and ethnic "cleansing"
- psychological, social, and economic instabilities
- civil wars and weapons trade
- the threat of nuclear proliferation and of the misuse of nuclear materials

Deterioration of the global situation

(see page 195)

The deterioration of the global situation continues rapidly with the immense growth in world population and in world production, severe damages to the environment and to biodiversity, a growing economic gap between the affluent and the poor, the dangers of civil war, the problems of "globalisation".

(See also the lists in chapter 22, pages 221-224)

d. Tasks for the institutions of science

Enlist psychology

(see page 1)

Psychology must be among the disciplines enlisted for an interdisciplinary investigation of various kinds of man-made or technology-made disasters, their origins, and their prevention.

The need for scientific analysis

(see page 2)

In order to avoid unpleasant surprises, a thorough and unbiased scientific analysis of the available options will be required.

A mission of the institutions of science

(see pages 3-4)

Humankind is faced with a twofold task of stupendous complexity:
1. to analyze the options available for action, particularly in economic and environmental policy, estimating for each option the risks, costs and benefits, taking into account by- and after-effects not only in the field of action chosen originally, but in neighbouring fields as well,
2. to study the history and psychology underlying the conflicts between neighbouring nations, and between ethnic and religious groups within nations, in time before war, or civil war, erupts, and to suggest joint tasks to be undertaken by the potential "enemies" which could make them understand that they are passengers in the same boat who depend on each other.
It is clear that the institutions of science (including psychology, anthropology and political science) and of letters (including history and economy) have an interdisciplinary and international mission here.

The effort required for the supply of interdisciplinary advice

(see pages 194, 196)

There can be no doubt that the world community will be in need of reliable, interdisciplinary scientific advice if major disasters are to be avoided. Because of the complexity of this task and the considerable cost involved in setting up a network of highly competent interdisciplinary working groups for this purpose, an extraordinary effort will be required. This should be the responsibility of the international community of the institutions of science, humanities and international law. There is no other international community or institution with comparable access to unbiased expertise, and at the same time with sufficient prestige to make itself heard.

By- and after-effects and multidimensional thinking

(see pages 3, 8, 28)

With the assistance of science and technology it would no doubt be feasible in many cases also to foresee the occurrence of by- and after-effects of human activities. They could be taken into account and made part of the overall planning. Countermeasures could be prepared, or the activities planned could be replaced by others with less harmful consequences. As the main effects and the by-effects often occur in different fields of specialization (in security policy and in psychology, for instance, or in economy and in climate research), multidimensional thinking and interdisciplinary collaboration are required.

Identify protection areas

(see page 9)

For species of animals and plants threatened by extinction sufficiently large protection areas and natural habitats have to be created.

Identify error intervals in risk and safety calculations, and limits to prognosticability

(see pages 9, 36/37)

It is not sufficient to set arbitrary safety factors in order to minimize the risk of accidents. It will rather be necessary to investigate all possibilities for errors in the behaviour of all members of all potential operation, maintenance and repair teams, taking into account their cultural backgrounds and psychological attitudes.

The limits to prognosticability, when too many parameters are involved, should also be clearly stated. At the same time attention should be called to the fact that certain (e.g. catastrophic) events can happen even if, due to the complexity of the situation, it is impossible to predict their occurrence with certainty. This must be taken into account in taking precautionary measures.

Prepare cost-benefit analyses

(see pages 9/10)

It is not the task of science to decide which risks society ought to take in order to reap which benefits. This is a decision based on values which can only be taken by the public, or by the politicians representing the public, after understanding the results of a thorough and carefully explained cost-benefit analysis.

Preservation of knowledge

(see page 13)

Essential prerequisite for avoiding, or at least limiting, the perils associated with a collapse of parts of the complex network of modern society, is preservation of the knowledge necessary for operating all components of the network. Not only technical know-how is required but also managerial, organisational, and psychological knowledge. Of great importance is the rooting of an ethical basis in a population from which the persons to be responsible for the control, further development and improvement of the network must be recruited. Accuracy, reliability, sense of duty are indispensable qualities of the "service personnel" for the machinery of modern society.

Need for international, interdisciplinary scientific committees and task forces

(see pages 20, 31, 37, 52, 80, 226)

Decision-makers need expert advice. This is a challenge for scientific institutions to form international interdisciplinary committees of scientists, engineers and scholars (including, if necessary, experts on psychology, international relations, international law, history, economy, arms control, political science, ecology, nutrition, geography, etc.) who devote themselves in an impartial way to the task of developing options for political action. For each option, the costs, risks and benefits should be stated, taking into account, if possible, short-term and long-term side- and after-effects.

Obviously, it will be necessary to define and select the "tasks" which the international task forces are to address in an interdisciplinary way. The *long-range dangers* (see above) threatening mankind are well known. A failure of the academic institutions to show concern and to offer advice will, in the long run, lead to the end of academic freedom as we know and cherish it. Scientists will be blamed for the risks which their work has produced for society unless they teach society how to master these risks in such a way that the benefit of their work outweighs by far the unavoidable negative aspects.

Standing international and interdisciplinary working groups could be set up in which leading experts constantly monitor developments causing concern because of their potentially dangerous trends and consequences. These working groups should elaborate options for action, giving estimates of chances, risks, and costs for each action.

The four levels of conflict resolution and of the avoidance of environmental catastrophes

(see pages 30, 36, 73)

The tasks are on four levels:

- the level of science and letters: the knowledge of the nature of latent and open conflicts, of their origins, and of the possibilities for settling them by negotiations, mediation, and by the application of international law, as well as the knowledge of the origin of environmental problems, still has many gaps which need to be filled by research,
- the level of practical policy: options for concrete action have to be worked out, with costs, benefits and risks assessed for each option,
- the level of the media: politicians and the public have to be informed about the situation so that the required political measures even if unpopular at the outset - get the necessary support,
- the level of education: the young generation has to be kept informed about the state of global affairs so that future leaders are prepared for the tasks they will have to tackle. Knowledge is not inherited, it must be taught and learned.

Scientists and scholars are able to, and therefore have a special responsibility, to work at all of these four levels.

Consider the obstacles

(see page 37)

What are the obstacles? What are the means by which they could possibly be overcome? What is the probable relation of costs to benefits in each case?

Need for structural adjustments?

(see pages 55/56)

All possible efforts must be directed towards finding ways of avoiding, or at least mitigating, disasters threatening peace. This needs knowledge, and to a large part scientific knowledge. The scientific institutions whose task it is to collect, screen, teach and preserve knowledge, are certainly called upon to make use of their resources for this purpose. This may need some structural adjustments. International and interdisciplinary working groups of experts carefully selected for each problem or task to be addressed, will have to be set up which permanently follow the developments in the fields of general concern (environmental pollution, population explosion, nuclear proliferation, arms trade, economic aspects of North-South and East-West relations, unemployment in industrial countries, psychological roots of civil wars and xenophobia, deficits in education etc). The institutions of science, of course, have no mandate to tell politicians and the public what should be done, but they are in a position to work out in detail what might happen unless certain measures are taken. They can also produce options for action, specifying for each option the risks, the potential benefits and the costs.

The first steps

(see page 56)

As a first step it will probably be necessary to take stock of the state of knowledge regarding the individual threats to human security in the various fields, and of their interconnections. At the same time a survey could be made of the existing recommendations by expert bodies in the fields of general concern mentioned above.

The next step would have to be to investigate for each of the problematic fields which obstacles are responsible for the fact that many of these recommendations are not implemented.

Dr. Vamik Volkan's method

(see pages 63, 66, 74)

Dr. Volkan and his colleagues have developed a kind of group intervention process in which leading representatives of the two opposing sides in a conflict meet with a neutral group of experts in psychoanalysis and in the history of the area in question. Conflicts between ethnic groups must be solved by explaining to the parties involved, through disinterested third parties and on the basis of all the historical, cultural and particularly psychological information available,[110]

- the origin of their conflict
- the reason why each party believes that its case is just
- the weaker points within these reasons as judged by neutral observers, and
  which possibilities exist for a compromise..

The prevention of armed conflicts could be based on a combination of Volkan's three-groups method with a definition of urgent joint goals. Meanwhile, long-range efforts should be devoted to the creation of international institutions of arbitration and peace enforcement which would be universally accepted and respected, except by a few trespassers who would be made to accept them.

The final programme

(see pages 72/73)

One might summarize the final programme in the following way:

- What are the great problems of our time which threaten the peaceful development of humankind, thereby creating fear, irrationality, violence and suffering? List them.
- What are the visible solutions? Are the available recommendations compatible with each other? List them.

---

[110]See, for instance: Methodology for reduction of ethnic tension, and promotion of democratization and institution building, Centre for the Study of Mind and Human Interaction, University of Virginia, March 1995.

- What are the obstacles preventing the implementation of a rational strategy for the solution of the problems? List them. How can the obstacles be overcome?

Use of modern communication technologies

(see pages 73, 75)

We can use the new possibilities of world-wide communication and of general access to the world's stored wisdom and knowledge for the definition and detailed description of humanity's common tasks for the next millennium, for the joint fight against regional and global threats to human beings and to nature, for the preservation of life on our planet. We have to feed the new systems with the information needed for the assessment of the available options.

Discuss modern European society and its role

(see page 83)

There is much that Europeans can learn from traditional societies. On the other hand, an honest and open discussion of the problems of modern European society, in contrast to those of African, Asian and Latin-American societies, could contribute towards mutual understanding and could assist the efforts towards the design of an equitable global system with a constitution enforcing the peaceful settlement of conflicts.

Scientists as experts and as citizens

(see pages 161/162)

Scientists should clearly distinguish between their statements, as experts, on what the facts are, and their statements, as politically interested citizens, on what should be done. Scientists, in their capacity as experts, are able to pronounce what should be done only if a clearly defined goal is given, like putting a man on the moon. They are also able to say, what should *not* be done if certain consequences are to be avoided, like heating up the earth's atmosphere, or creating hatred between peoples.

The requirement of global co-ordination

(see page 165)

Global co-ordination is needed if environmental and other catastrophes of a global character are to be avoided in the years to come. This means that the United Nations must be strengthened, and that interdisciplinary scientific advice of top quality must be made available to decisionmakers at the global level. In this context it is of interest to take stock of the existing advisory mechanisms of the United Nations. Although they are mostly not coordinated, usually serving one single agency and its missions, they play a useful role and could serve as contributors to, or "building blocks" of, an overall interdisciplinary, electronically connected, advisory system for the avoidance of future catastrophes.

Climate research

(see pages 193/194)

"The development of an optimal, internationally agreed upon, strategy for climate protection is a very complex task of contemporary politics. Such a strategy would have to take into account the costs of adaptation (to a potential global rise of temperature) as well as those of avoiding changes of climate by restructuring industry. In order to support the efforts to fulfill this task, science should supply ground-data as reliable as possible. This requires a close collaboration between climatologists, ecologists, economists, legal experts and representatives of other disciplines. ... Scientifically unfounded 'opinions', whether by prophets of an approaching climate catastrophe, or by climate skeptics, do not bring any progress." (Klaus Hasselmann)

Appendix

About the author

Klaus L.F. Gottstein (born 1924 in Stettin, Germany)

Group Leader (1958-65), Division Leader (1965-71), one of the directors of the Max Planck Institute for Physics, Munich (1971-83)

Professor at the University of Munich (1967-)

Counsellor for Scientific Affairs (Science Attaché) at the Embassy of the Federal Republic of Germany (FRG) in Washington, D.C. (1971-74)

Head of the delegation of the FRG to the meeting of experts of the member countries of the Conference on Security and Co-operation in Europe (CSCE) convened to prepare a Scientific Forum of the CSCE (Bonn, 1978)

Coordinator of scientific contributions of the FRG to the United Nations Conference on Science and Technology for Development (Vienna, 1979)

Executive Secretary of the Scientific Forum of the CSCE (Hamburg, 1980)

Speaker of the Pugwash Group of the FRG (1976-87), organizer of the Quinquennial Pugwash Conference on Science and World Affairs (Munich, 1977), and participant in all 23 annual Pugwash Conferences from 1975 to 1997

Member of the advisory committee for the International Institute for Applied Systems Analysis, IIASA, Laxenburg (1975-84)

Member of the Advisory Committee on Science, Technology and Society to UNESCO (Paris, 1981-83)

Director, Research Unit (on problems of science and society) in the Max-Planck-Gesellschaft (Munich, 1984-92)

Organizer of international conferences and workshops on the "Strategic Defense Initiative" (1986) and on the mutual perceptions of long-range goals in East-West relations and in European re-orientation (1987, 1989, 1990, 1991)

Co-organizer of the International Amaldi Conference "International Security in a Transformed World" (Heidelberg, 1992)

Member of Advisory Panel on Science Policy to the Ministry of Science of the Russian Federation, Moscow (1993)

Consultant to Academy of Sciences of the Republic of Moldova under the auspices of the United Nations Development Programme, UNDP, Chisinau (1994)

Representative of the Union of German Academies of Sciences and Letters with the International Amaldi Conferences of Academies of Sciences and National Scientific Societies on Problems of Global Security (1998)

Index